Algorithm and Design Complexity

Computational complexity is critical in analysis of algorithms and is important to be able to select algorithms for efficiency and solvability. *Algorithm and Design Complexity* initiates with discussion of algorithm analysis, time–space trade-off, symptotic notations, and so forth. It further includes algorithms that are definite and effective, known as computational procedures. Further topics explored include divide-and-conquer, dynamic programming, and backtracking.

Features:

- Includes complete coverage of basics and design of algorithms
- Discusses algorithm analysis techniques like divide-and-conquer, dynamic programming, and greedy heuristics
- Provides time and space complexity tutorials
- Reviews combinatorial optimization of Knapsack problem
- Simplifies recurrence relation for time complexity

This book is aimed at graduate students and researchers in computers science, information technology, and electrical engineering.

Algorithm and Design Complexity

Anli Sherine, Mary Jasmine,
Geno Peter, and S. Albert Alexander

CRC Press
Taylor & Francis Group
Boca Raton London New York

CRC Press is an imprint of the
Taylor & Francis Group, an **informa** business

Designed cover image: Shutterstock

First edition published 2023
by CRC Press
6000 Broken Sound Parkway NW, Suite 300, Boca Raton, FL 33487–2742

and by CRC Press
4 Park Square, Milton Park, Abingdon, Oxon, OX14 4RN

CRC Press is an imprint of Taylor & Francis Group, LLC

ISBN: 978-1-032-40932-0 (hbk)
ISBN: 978-1-032-40935-1 (pbk)
ISBN: 978-1-003-35540-3 (ebk)

DOI: 10.1201/9781003355403

Typeset in Times
by Apex CoVantage, LLC

Contents

Preface

Algorithms have been an idea since ancient times. Ancient mathematicians in Babylonia and Egypt used arithmetic methods, such as a division algorithm, around 2500 BCE and 1550 BCE, respectively. Later, in 240 BCE, Greek mathematicians utilized algorithms to locate prime numbers using the Eratosthenes sieve and determine the greatest common divisor of two integers using the Euclidean algorithm. Al-Kindi and other Arabic mathematicians of the ninth century employed frequency-based cryptography techniques to decipher codes. Both the science and the practice of computers are centered on algorithms. Since this truth has been acknowledged, numerous textbooks on the topic have come to be. Generally speaking, they present algorithms in one of two ways. One categorizes algorithms based on a certain problem category. The three main objectives of this book are to raise awareness of the impact that algorithms can have on the effectiveness of a program, enhance algorithm design skills, and develop the abilities required to analyze any algorithms that are utilized in programs. Today's commercial goods give the impression that some software developers don't give space and time efficiency any thought. They anticipate that if a program uses too much memory, the user will purchase additional memory. They anticipate that if an application takes too long, the customer will get a faster machine.

The emphasis on algorithm design techniques is due to three main factors. First off, using these strategies gives a pupil the means to create algorithms for brand-new issues. As a result, studying algorithm design approaches is a highly beneficial activity. Second, they attempt to categorize numerous existing algorithms in accordance with a fundamental design principle. One of the main objectives of computer science education should be to teach students to recognize such similarities among algorithms from various application domains. After all, every science views the classification of its main topic as a major, if not the discipline's focal point. Third, we believe that techniques for designing algorithms are useful as generic approaches to solving issues that transcend beyond those related to computing. There are a number of significant issues, both theoretically and educationally. This book is intended as a manual on algorithm design, providing access to combinatorial algorithm technology for both student and computer professionals.

Anli Sherine
Mary Jasmine
Geno Peter
S. Albert Alexander

Acknowledgments

First and foremost, praises and thanks to the God, the Almighty, for his showers of blessings that helped us in prewriting, research, drafting, revising, editing, proofreading, and, finally, a successful book to share the knowledge.

"Being deeply loved by someone gives you strength, while loving someone deeply gives you courage." Anli Sherine and Dr. Geno Peter would like to express our love for each other. We wish to thank our sweethearts, Hadriel Peter and Hanne Geona, for giving us the love and space during the writing process.

Mary Jasmine wishes to thank all his lovable and sweet family members. They have given me a lot of support and encouragement to complete this book.

Dr. S. Albert Alexander would like to take this opportunity to acknowledge those people who helped me in completing this book. I am thankful to all my research scholars and students who are doing their project and research work with me. But the writing of this book is possible mainly because of the support of my family members, parents, and brothers. Most important, I am very grateful to my wife, A. Lincy Annet, for her constant support during writing. Without her, all these things would not be possible. I would like to express my special gratitude to my son, A. Albin Emmanuel, for his smiling face and support; it helped a lot in completing this work.

About the Authors

Anli Sherine graduated with a Bachelor of Technology in information technology from Anna University, India, and subsequently completed her Master of Engineering in computer science engineering from Anna University, India. Currently, she works in the School of Computing and Creative Media of the University of Technology Sarawak, Malaysia. She is a member of the Malaysian Board of Technologists. Her research interests include, but are not limited to, cryptography, mobile computing, and digital image processing.

Mary Jasmine is currently an assistant professor in the Department of Computer Science and Engineering at Dayananda Sagar University, India. She received her Master of Engineering in computer science engineering from Anna University, India. She received her Bachelor of Engineering in computer science engineering from Anna University, India. Her research interest is in the area of machine learning techniques for big data analytics and its applications.

Geno Peter graduated with a Bachelor of Engineering in electrical and electronics engineering from Bharathiar University, India; subsequently completed a Master of Engineering in power electronics and drives from Karunya University, India; and then received a Doctor of Philosophy in electrical engineering from Anna University, India. He started his career as a test engineer with General Electric (a transformer manufacturing company) in India; subsequently worked with Emirates Transformer & Switchgear, in Dubai, as a test engineer; and then worked with Al-Ahleia Switchgear Company, in Kuwait, as a quality assurance engineer. He is a trained person to work on HAEFELY, an impulse testing system, in Switzerland. He is a trained person to work on Morgan Schaffer, a dissolved gas analyzer testing system, in Canada. His research interests are in transformers, power electronics, power systems, and switchgears. He has trained engineers from the Government Electricity Board in India on testing various transformers. He has given hands-on training to engineers from different oil and gas companies in Dubai and Kuwait on testing transformers and switchgears. He has published his research findings in 41 international and national journals. He has presented his research findings in 17 international conferences. He is the author of the book *A Typical Switchgear Assembly*. He is a Chartered Engineer and a Professional Engineer of the Institution of Engineers (India).

S. Albert Alexander was a postdoctoral research fellow from Northeastern University, Boston, Massachusetts, USA. He is the recipient of the prestigious Raman Research Fellowship from the University Grants Commission (Government of India). His current research focuses on fault diagnostic systems for solar energy conversion systems and smart grids. He has 15 years of academic and research experience. He has published 45 technical papers in international and national journals (including *IEEE Transactions* and Institution of Engineering and Technology (IET) and those published by Elsevier, Taylor & Francis, and Wiley, among others) and presented 45 papers at national and international conferences. He has completed four Government of India–funded projects, and three projects are under progress, with the overall grant amount of Rs. 2.3 crores. His PhD work on power quality earned him a National Award from the Indian Society for Technical Education (ISTE), and he has received 23 awards for his meritorious academic and research career (such as Young Engineers Award from IE(I), Young Scientist Award from Sardar Patel Renewable Energy Research Institute (SPRERI), Gujarat, among others). He has also received the National Teaching Innovator Award from the Ministry of Human Resource Development (MHRD) (Government of India). He is an approved "Margadarshak" from All India Council for Technical Education (AICTE) (Government of India). He is the approved Mentor for Change under the Atal Innovation Mission. He has guided 35 graduate and postgraduate projects. He is presently guiding six research scholars, and five have completed their PhDs. He is a member and in prestigious positions in various national and international forums (such as senior member of IEEE and vice president for the Energy Conservation Society, India). He has been an invited speaker in 220 programs covering nine Indian states and in the United States. He has organized 11 events, including faculty development programs, workshops, and seminars. He completed his graduate program in electrical and electronics engineering at Bharathiar University and his postgraduate program at Anna University, India. Presently he is working as an associate professor with the School of Electrical Engineering, Vellore Institute of Technology, India, and is doing research work on smart grids, solar photovoltaic (PV), and power quality improvement techniques. He has authored several books in his areas of interest.

Introduction

The complexity of a problem is the complexity of the best algorithms that allow solving the problem. The study of the complexity of explicitly given algorithms is called analysis of algorithms, while the study of the complexity of problems is called computational complexity theory.

In Chapter 1 we discuss algorithm analysis, time–space trade-off, symptotic notations, properties of big-oh notation, conditional asymptotic notation, recurrence equations, solving recurrence equations, and analysis of a linear search. In Chapter 2, we discuss in detail divide and conquer: the general method, binary search, finding the maximum and minimum, merge sort, and greedy algorithms: the general method, container loading, and the knapsack problem. Chapter 3 talks about dynamic programming: the general method, multistage graphs, all-pair shortest paths, optimal binary search trees, the 0/1 knapsack problem, and the traveling salesperson problem. Chapter 4 talks about backtracking: the general method, the 8-queens problem, the sum of subsets, graph coloring, the Hamiltonian problem, and the knapsack problem. The final chapter talks about graph traversals, connected components, spanning trees, biconnected components, branch and bound: general methods (first in, first out and least cost) and the 0/1 knapsack problem, and an introduction to NP-hard and NP completeness.

We discuss algorithms that are definite and effective, also called computational procedures. Both areas are highly related, as the complexity of an algorithm is always an upper bound on the complexity of the problem solved by this algorithm. Moreover, for designing efficient algorithms, it is often fundamental to compare the complexity of a specific algorithm to the complexity of the problem to be solved. Also, in most cases, the only thing that is known about the complexity of a problem is that it is lower than the complexity of the most efficient known algorithms. Therefore, there is a large overlap between the analysis of algorithms and complexity theory. In computer science, the computational complexity, or simply complexity, of an algorithm is the number of resources required to run it. Computational complexity is very important in the analysis of algorithms. As problems become more complex and increase in size, it is important to be able to select algorithms for efficiency and solvability. The ability to classify algorithms based on their complexity is very useful.

1 Algorithm Analysis

1.1 ALGORITHM ANALYSIS

Why do you need to study algorithms? If you are going to be a computer professional, there are both practical and theoretical reasons to study algorithms. From a practical standpoint, you have to know a standard set of important algorithms from different areas of computing; in addition, you should be able to design new algorithms and analyze their efficiency. From a theoretical standpoint, the study of algorithms, sometimes called *algorithmics*, has come to be recognized as the cornerstone of computer science.

Another reason for studying algorithms is their usefulness in developing analytical skills. After all, algorithms can be seen as special kinds of solutions to problems— not just answers but precisely defined procedures for getting answers. Consequently, specific algorithm design techniques can be interpreted as problem-solving strategies that can be useful regardless of whether a computer is involved. Of course, the precision inherently imposed by algorithmic thinking limits the kinds of problems that can be solved with an algorithm.

There are many algorithms that can solve a given problem. They will have different characteristics that will determine how efficiently each will operate. When we analyze an algorithm, we first have to show that the algorithm does properly solve the problem because if it doesn't, its efficiency is not important. Analyzing an algorithm determines the amount of 'time' that an algorithm takes to execute. This is not really a number of seconds or any other clock measurement but rather an approximation of the number of operations that an algorithm performs. The number of operations is related to the execution time, so we will sometimes use the word *time* to describe an algorithm's computational complexity. The actual number of seconds it takes an algorithm to execute on a computer is not useful in our analysis because we are concerned with the relative efficiency of algorithms that solve a particular problem.

WHAT IS AN ALGORITHM?

The word *algorithm* comes from the name of a Persian author Abu Ja'far Muhammad ibn Musa al-Khwarizmi who wrote a textbook on mathematics. This word has taken on a special significance in computer science, where 'algorithm' has come to refer to a method that can be used by a computer for the solution of a problem.

Definition: An algorithm is a finite set of instructions that accomplishes a particular task.

All algorithms must satisfy the following criteria:

1. **Input**. Zero or more quantities are externally supplied.
2. **Output**. At least one quantity is produced.

DOI: 10.1201/9781003355403-1

3. **Definiteness**. Each instruction is clear and unambiguous.
4. **Finiteness**. If we trace out the instructions of an algorithm, then for all cases, the algorithm terminates after a finite number of steps.
5. **Effectiveness**. Every instruction must be very basic. It also must be feasible. An algorithm is composed of a finite set of steps, each of which may require one or more operations.

Computational procedures: Algorithms that are definite and effective are also called computational procedures. One important example of computational procedures is the operating system of a digital computer. This procedure is designed to control the execution of jobs in such a way that when no jobs are available, it does not terminate but continues in a waiting state until a new job is entered.

Notion of an algorithm: An algorithm is a sequence of unambiguous instructions for solving a problem, that is, for obtaining a required output for any legitimate input in a finite amount of time as shown in Figure 1.1.

To illustrate the notion of algorithm let us consider three methods for solving the same problem. The three methods are

1. Euclid's algorithm,
2. consecutive integer checking algorithm, and
3. the middle school procedure.

1. Euclid's Algorithm

Euclid of Alexandria (third century BCE) outlined an algorithm for solving this problem in one of the volumes of his *Elements*, most famous for its systematic exposition of geometry. In modern terms, ***Euclid's algorithm*** is based on applying repeatedly the equality

$$\gcd(m, n) = \gcd(n, m \bmod n)$$

where *m mod n* is the remainder of the division of *m* by *n*, until *m* mod *n* is equal to 0. The last value of *m* is also the greatest common divisor (gcd) of the initial *m* mod *n*. It is used to compute gcd of two integers in an effective manner.

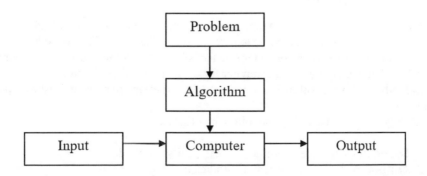

FIGURE 1.1   The notion of the algorithm

Example 1
$$gcd(60, 24) = gcd(24, 60 \bmod 24)$$
$$= gcd(24, 12)$$
$$= gcd(12, 24 \bmod 12)$$
$$= gcd(12, 0)$$
$$= 12$$

Euclid's algorithm for computing gcd (m, n):

Step 1: If $n = 0$, return the value of m as the answer and stop; otherwise, proceed to step 2.
Step 2: Divide m by n and assign the value of the remainder to r.
Step 3: Assign the value of n to m and the value of r to n. Go to step 1.

Algorithm (Euclid(m,n)) in a pseudo code:

```
// Compute gcd (m,n) by Euclid's algorithm
// Input: Two nonnegative and nonzero integers m and n
//Output: Greatest common divisor of m and n.
while n ≠ 0 do
    r ← m mod n
    m ‹ n
    n ←r
return m
```

How do we know that Euclid's algorithm eventually comes to a stop? This follows from the observation that the second integer of the pair gets smaller with each iteration and that it cannot become negative. Indeed, the new value of n on the next iteration is $m \bmod n$, which is always smaller than n (why?). Hence, the value of the second integer eventually becomes 0, and the algorithm stops.

2. Consecutive Integer Checking Algorithm for Computing gcd(m, n)
A common divisor cannot be greater than the smaller of the two numbers, which will denote by

$$t = \min \{m, n\}.$$

So we can check whether t divides both m and n. If it does t is the answer; If it does not, we simply decrease t by 1 and try again. We have to follow this method until it reaches the answer.

Step 1: Assign the value of $\min\{m, n\}$ to t.
Step 2: Divide m by t. If the remainder of this division is 0 go to step 3; otherwise, go to step 4.
Step 3: Divide n by t. If the remainder of this division is 0, return the value of t as the answer and stop; otherwise, proceed to step 4.
Step 4: Decrease the value of t by 1. Go to step 2.

Let us understand this algorithm with the help of some examples.

Consider $m = 12$, $n = 8$
$t = \min(12, 8)$
We will set a value of $t = 8$ initially.

Check whether we get $m \bmod t = 0$ as well as $n \bmod t = 0$.
If not, then decrease t by 1, and again, with this new t value, check whether

$m \bmod t = 0$ and $n \bmod t = 0$.

Then we go on checking whether $m \bmod t$ and $n \bmod t$ both result in 0 or not. Thus, we will repeat this process each time by decrementing t by 1 and performing $m \bmod t$ and $n \bmod t$.

$12 \bmod 8 = 4$	$8 \bmod 8 = 0$	12 mod 8 is not equal to zero. So 8 is not a gcd.
So set $t = t - 1$	that is, $t = 8 - 1 = 7$	
$12 \bmod 7 = 5$	$8 \bmod 7 = 1$	Both 12 mod 7 and 8 mod 7 are not equal to 0. So 7 is not a gcd. So set new t $= t - 1$, that is, $t = 7 - 1 = 6$.
$12 \bmod 6 = 0$	$8 \bmod 6 = 2$	8 mod 6 is not equal to zero. So 6 is not a gcd.
So $t = 6 - 1 = 5$.		
$12 \bmod 5 = 7$	$8 \bmod 5 = 3$	Both 12 mod 5 and 8 mod 5 are not equal to 0. So 5 is not a gcd. So set new $t = t - 1$, that is, $t = 5 - 1 = 4$.
$12 \bmod 4 = 0$	$8 \bmod 4 = 0$	Both 12 mod 4 and 8 mod 4 are equal to 0. So 4 is a gcd.
$\gcd(12, 8) = 4$		

3. Middle School Procedure for Computing gcd(m, n)

Step 1: Find the prime factors of m.
Step 2: Find the prime factors of n.
Step 3: Identify all the common factors in the two prime expansions found in steps 1 and 2.
Step 4: Compute the product of all the common factors and return it as the gcd of the numbers given.

Thus, for the numbers 120 and 72, we get

$$120 = 2 \times 2 \times 2 \times 3 \times 5$$
$$72 = 2 \times 2 \times 2 \times 3 \times 3$$
$$\gcd(120, 72) = 2 \times 2 \times 2 \times 3 = 24.$$

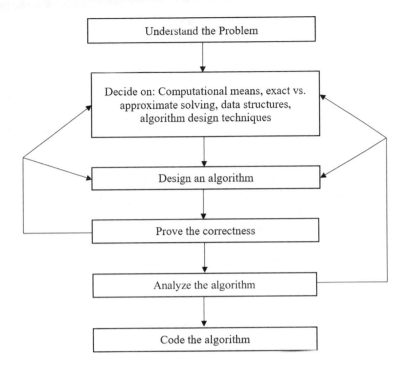

FIGURE 1.2   Algorithm design and analysis process

FUNDAMENTALS OF ALGORITHM PROBLEM-SOLVING

In computer science, developing an algorithm is an art or skill. Before the actual implementation of the program, designing an algorithm is a very important step. We can consider algorithms to be procedural solutions to problems. These solutions are not answers but specific instructions for getting answers. It is this emphasis on precisely defined constructive procedures that makes computer science distinct from other disciplines. In particular, this distinguishes it from theoretical mathematics, whose practitioners are typically satisfied with just proving the existence of a solution to a problem and, possibly, investigating the solution's properties. We now list and briefly discuss a sequence of steps one typically goes through when designing and analyzing an algorithm as shown in Figure 1.2.

A Sequence of Steps in Designing and Analyzing an Algorithm
- Understanding the problem
- Ascertaining the capabilities of a computational device
- Choosing between exact and approximate problem-solving
- Deciding the appropriate data structures
- Reviewing algorithm design techniques
- Designing the algorithm

- Proving the correctness of the algorithm
- Analyzing the algorithm
- Coding the algorithm

Understanding the Problem

- From a practical perspective, first of all, the step we need to understand completely is the problem statement. To understand the problem statements, read the problem description carefully and ask questions to clarify the doubts about the problem.
- After understanding the problem statements, find out what the necessary inputs for solving that problem are. Use a known algorithm for solving it.
- If there is no readily available algorithm, then design your own.
- An input to the algorithm is called the instance of the problem. It is very important to decide the range of inputs so that the boundary values of an algorithm get fixed.
- The algorithm should work correctly for all valid inputs.

Ascertaining the Capabilities of a Computational Device

Once you completely understand a problem, you need to ascertain the capabilities of the computational device the algorithm is intended for. We can classify an algorithm according to its execution point of view as a sequential algorithm and a parallel algorithm.

- **Sequential algorithm:** The sequential algorithm runs on the machine in which the instructions are executed one after another. Such a machine is called a random-access machine. Algorithms designed to be executed on such machines are called sequential algorithms.
- **Parallel algorithm:** Some newer computers can execute operations concurrently, that is, in parallel. Algorithms designed to be executed on machines that execute operations concurrently are known as parallel algorithms.

Choosing Between Exact and Approximate Problem-Solving

The next decision is to choose between solving the problem exactly or solving the problem approximately.

- An algorithm that solves the problem exactly is called an **exact algorithm**.
- An algorithm that solves the problem approximately is called an **approximation algorithm**.

Reasons for Using an Approximation Algorithm

- There are important problems that cannot be solved exactly.
- Available algorithms for solving a problem exactly can be unacceptably slow because of the problem's intrinsic complexity.
- An approximation algorithm can be a part of a more sophisticated algorithm that solves a problem exactly.

Deciding on appropriate data structures: The efficiency of an algorithm can be improved by using an appropriate data structure. The data structure is important for both the design and the analysis of an algorithm. The data structure and the algorithm work together, and these are interdependent. Hence, choosing the proper data structure is required before designing the actual algorithm.

Algorithm design techniques: An algorithm design technique is a general approach to solve problems algorithmically. These problems may belong to different areas of computing.

Various algorithm design techniques are

- **Divide and conquer:** In this strategy, the problem is divided into smaller sub-problems; the subproblems are solved to obtain the solution to the main problem.
- **Dynamic programming:** The problem is divided into smaller instances, and the results of smaller reoccurring instances are obtained to solve the problem.
- **Greedy technique:** From a set of obtained solutions, the best possible solution is chosen each time to obtain the final solution.
- **Backtracking:** In this method, in order to obtain the solution, a trial-and-error method is followed.

Methods of Specifying an Algorithm

Once you have designed an algorithm, you need to specify it in some fashion.

There are different ways by which we can specify an algorithm:

1. Using natural language
2. Pseudo code
3. Flowchart

1. **Using natural language:** It is very simple to specify an algorithm using natural language.
   For example: Write an algorithm to perform the addition of two numbers.
   Step 1: Read the first number, say, 'a'.
   Step 2: Read the second number, say, 'b'.
   Step 3: Add the two numbers and store the result in a variable.
   Step 4: Display the result.
2. **Pseudo code:** In pseudo code, English-like words are used to represent the various logical steps. The pseudo code is a mixture of natural language and programming language–like constructs.

   For example:
   ALGORITHM Sum(a,b)
   //Problem Description: This algorithm performs addition of two numbers
   //Input: Two integers a and b
   // Output: Addition of two integers

   $$c \leftarrow a + b$$

   return c.
3. **Flowchart:** A flowchart is a visual representation of the sequence of steps for solving a problem. Instead of descriptive steps, we use pictorial

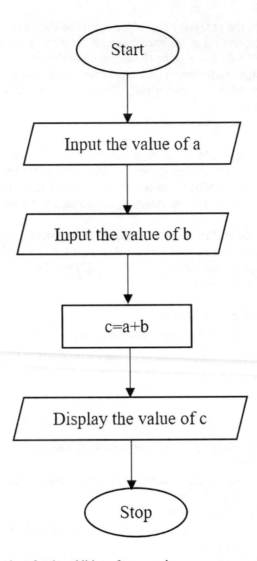

FIGURE 1.3   Flowchart for the addition of two numbers

representation for every step. A flowchart is a set of symbols that indicates various operations in the program. For example, a flowchart for the addition of two numbers is shown in Figure 1.3.

Proving the Correctness of the Algorithm
- Once an algorithm has been specified, you have to prove its correctness. That is, you have to prove that the algorithm yields a required result for every legitimate input in a finite amount of time.

- A common technique for proving correctness is to use mathematical induction because an algorithm's iterations provide a natural sequence of steps needed for such proofs.
- But in order to show that an algorithm is incorrect, you need just one instance of its input for which the algorithm fails.
- If the algorithm is found to be incorrect, you need to redesign it under the same decisions regarding the data structures, the design techniques, and so on.

Analyzing the Algorithm

After correctness, by far the most important is efficiency. While analyzing an algorithm, we should consider the following factors:

- Time efficiency
- Space efficiency
- Simplicity
- Generality

Time efficiency: Time efficiency indicates how fast the algorithm runs.
Space efficiency: Space efficiency indicates how much extra memory the algorithm needs to complete its execution.
Simplicity: Simplicity of an algorithm means generating a sequence of instructions that are easy to understand.
Generality: The generality of the problem is what the algorithm solves and the range of inputs it accepts.

If you are not satisfied with the algorithm's efficiency, simplicity, or generality, you must return to the drawing board and redesign the algorithm.

Coding the Algorithm

The implementation of an algorithm is done through a suitable programming language.

The validity of programs is still established by testing. Test and debug your program thoroughly whenever you implement an algorithm.

STUDY OF ALGORITHMS

The study of algorithms includes many important areas, such as

- how to devise algorithms,
- how to validate algorithms,
- how to analyze algorithms, and
- how to test a program.

How to devise algorithms: Creating an algorithm is an art. By mastering the design strategies, it will become easier to devise new and useful algorithms.

How to validate algorithms: Once an algorithm is devised, it is necessary to show that it computes the correct answer for all possible legal inputs. This process is algorithm validation. The purpose of the validation is to assure that this algorithm will work correctly. Once the validity of the method has been shown, a program can be written, and a second phase begins. This phase is referred to as *program proving* or sometimes as *program verification*.

A proof of correctness requires that the solution be stated in two forms. One form is usually a program that is annotated by a set of assertions about the input and output variables of the program. These assertions are often expressed in the predicate calculus. The second form is called a *specification*, and this may also be expressed in the predicate calculus.

How to analyze algorithms: This field of study is called the analysis of algorithms. As an algorithm is executed it uses the computer's central processing unit to perform operations and its memory to hold the program and data. An analysis of algorithms, or a performance analysis, refers to the task of determining how much computing time and storage an algorithm requires.

How to test a program: Testing a program consists of two phases:

- Debugging
- Profiling

Debugging is the process of executing programs on sample data sets to determine whether faulty results occur and, if so, to correct them.

Profiling or Performance measurement is the process of executing a correct program on data sets and measuring the time and space it takes to compute the results.

IMPORTANT PROBLEM TYPES

Let us see some of the most important problem types:

1. Sorting
2. Searching
3. String processing
4. Graph problems
5. Combinatorial problems
6. Geometric problems
7. Numerical problems

1. **Sorting:** Sorting is nothing but arranging a set of elements in increasing or decreasing order. The sorting can be done on numbers, characters, strings, or employee records. As a practical matter, we usually need to sort lists of numbers, characters from an alphabet, character strings, and, most important, records similar to those maintained by schools about their students, libraries about their holdings, and companies about their employees. In the case of records, we need to choose a piece of information to guide sorting. For example, we can choose to sort student records

in alphabetical order of names or by student number or by student grade-point average. Such a specially chosen piece of information is called a **key**. Computer scientists often talk about sorting a list of keys even when the list's items are not records but, say, just integers. Furthermore, sorting makes many questions about the list easier to answer. The most important use of sorting is searching; it is why dictionaries, telephone books, class lists, and so on are sorted.

2. **Searching:** Searching is an activity by which we can find out the desired element from the list. The element which is to be searched is called a **search key**. There are many searching algorithms such as sequential search, binary search, and many more. They range from the straightforward sequential search to a spectacularly efficient but limited binary search and algorithms based on representing the underlying set in a different form more conducive to searching.

For searching, there is no single algorithm that fits all situations best. Some algorithms work faster than others but require more memory. Some are very fast but applicable only to sorted arrays. Unlike sorting algorithms, there is no stability problem, but different issues arise. Specifically, in applications in which the underlying data may change frequently relative to the number of searches, searching has to be considered in conjunction with two other operations: an addition to and deletion from the data set of an item. In such situations, data structures and algorithms should be chosen to strike a balance among the requirements of each operation. Also, organizing very large data sets for efficient searching poses special challenges with important implications for real-world applications.

3. **String processing:** A string is a sequence of characters from an alphabet. Strings of particular interrupt are text strings, which comprise letters, numbers, and special characters, and bit strings, which comprise zeros and ones. It should be pointed out that string-processing algorithms have been important for computer science for a long time in conjunction with computer languages and compiling issues.

4. **Graph problems:** A graph can be thought of as a collection of points called vertices, some of which are connected by line segments called edges. Basic graph algorithms include graph traversal algorithms, shortest path algorithms, and topological sorting for graphs with directed edges. Graphs are an interesting subject to study, for both theoretical and practical reasons. Graphs can be used for modeling a wide variety of applications, including transportation, communication, social and economic networks, project scheduling, and games. Studying different technical and social aspects of the internet in particular is one of the active areas of current research involving computer scientists, economists, and social scientists.

5. **Combinatorial problems:** Combinatorial problems are related to problems like computing permutations and combinations. Combinatorial problems are the most difficult problems in computing areas because of the following causes:

- As problem size grows, the combinatorial objects grow rapidly and reach to a huge value.
- There is no algorithm available that can solve these problems in a finite amount of time.
- Many of these problems fall in the category of unsolvable problems.
6. **Geometric problems:** Geometric algorithms deal with geometric objects such as points, lines, and polygons. Today, people are interested in geometric algorithms with quite different applications such as applications to computer graphics, robotics, and topography.
7. **Numerical problems:** Numerical problems are problems that involve mathematical objects of continuous nature, solving equations and systems of equations, computing definite integrals, evaluating functions, and so on. A majority of such mathematical problems can be solved only approximately.

ABSTRACT DATA TYPES

- The abstract data type (ADT) is a mathematical model which gives a set of utilities available to the user but never states the details of its implementation.
- In object-oriented programming, the class is an ADT, which consists of data and functions that operate on the data.
- Various visibility labels in object-oriented languages are used to restrict data access only through the functions of the object. That is, the data are secured by hiding the data.

Difference Between ADT and Data Types

Data type is an implementation or computer representation of an ADT. Once the ADT is defined, programmers can use the properties of the ADT by creating instances of the ADT. For example, a programming language provides several built-in data types.

Example

Integer is an ADT. The implementation of an *integer* may be through any one of the several forms like unsigned, signed, and others. The instance of this is used in programs. In C++, the instance of the data type *int* is created by declaring

  int i;

Here, the programmer simply uses the properties of the *int* by creating an instance without seeing how it has been implemented. Therefore, *int* can be said to be an INTEGER ADT.

- We cannot always expect all necessary ADTs to be available in the form of built-in data types. Sometimes, the user may want to represent the data type, both logically and physically, in a specified manner. This is the concept of the user-defined data type.

- In many situations, communication between data structures becomes mandatory, such as restricting one data structure to access the data of other without its knowledge.
- In programming languages, the concepts of classes permit this by communicating through the member functions.
- Structured programming languages do not have the properties of data hiding. This creates damage to the data.

Example

As an example, let us take the data structure **stack**. The stack is implemented using arrays, then, it is possible to modify any data in the array without going through the proper rule, LIFO (last in, first out). Figure 1.4 demonstrates this. The data's of this kind is not possible in an object-oriented approach. Figure 1.4 shows the direct access of data in the non-object-oriented approach. This is an illegal operation, but no program error occurs.

Performance Analysis: The analysis of an algorithm deals with the amount of time and space consumed by it. An efficient algorithm can be computed with minimum requirement of time and space.

Space Complexity:Most of what we will be discussing is going to be how efficient various algorithms are in terms of time, but some forms of analysis could be done based on how much space an algorithm needs to complete its task. This space complexity analysis was critical in the early days of computing when storage space on a computer (both internal and external) was limited. When considering space complexity, algorithms are divided into those that need extra space to do their work and those that work in place. It was not unusual for programmers to choose an algorithm that was slower just because it worked in place, because there was not enough extra memory for a faster algorithm. The space complexity of an algorithm is the amount of memory it needs to run to completion.

Looking at software on the market today, it is easy to see that space analysis is not being done. Programs, even simple ones, regularly quote space needs in a number of megabytes. Software companies seem to feel that making their software space efficient is not a consideration because customers who don't have enough computer

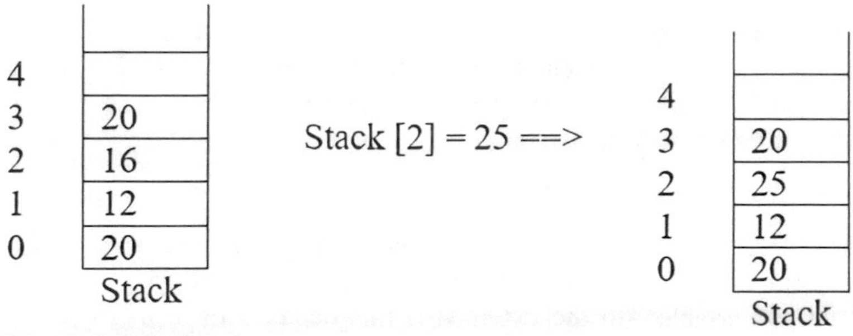

FIGURE 1.4   Example of a stack

memory can just go out and buy another 32 megabytes (or more) of memory to run the program or a bigger hard disk to store it. This attitude drives computers into obsolescence long before they really are obsolete.

Reasons for the Space Complexity of a Program

- If the program is to be run on a multi-user computer system, then the amount of memory to be allocated to the program needs to be specified.
- For any computer system, know in advance whether there is sufficient memory is available to run the program.
- A problem might have several possible solutions with different space requirements.
- Use the space complexity to estimate the size of the largest problem that a program can solve.

Components of Space Complexity

The space needed by each algorithm is the sum of the following components:

1. Instruction space
2. Data space
3. Environment stack space

1. *Instruction space.* The space needed to store the compiled version of the program instructions
2. *Data space:* The space needed to store all constant and variable values
3. *Environment stack space:* The space needed to store information to resume execution of partially completed functions

The total space needed by an algorithm can be simply divided into two parts from the *three components of space complexity*:

1. Fixed
2. Variable

Fixed Part

A fixed part is independent of the characteristics (e.g., number, size) of the inputs and outputs. This part typically includes the instruction space (i.e., space for the code), space for simple variables and fixed-size component variables (also called *aggregate*), space for constants, and so on.

Variable Part

A variable part that consists of the space needed by component variables whose size is dependent on the particular problem instance being solved, the space needed by referenced variables (to the extent that this depends on instance characteristics), and the recursion stack space (insofar as this space depends on the instance characteristics).

The space requirement **S(P)** of any algorithm **P** may therefore be written as

S(P) = c + Sp(instance characteristics), where c is a constant that denotes the
fixed part of the space requirement.

Sp—This variable component depends on the magnitude (size) of the inputs to
and outputs from the algorithm.

When analyzing the space complexity of a program, we concentrate solely
on estimating **Sp(instance characteristics)**. For any given problem, we need
first to determine which instance characteristics to use to measure the space
requirements.

Examples
Find the space complexity of the following algorithms:

1. Algorithm abc computes $a + b + b \times c + 4.0$.
   Algorithm abc (a,b,c)
   {
   return a+b+b x c+ 4.0;
   }

For algorithm abc, the problem instance is characterized by the specific values of
a, b, and c. Assume that one word is adequate to store the values of each a, b, and c,
and the result, which is the space needed by *abc*, is independent of the instant characteristics, **Sp**(instance characteristics) = 0.

2. Algorithm abc computes $a + b + b * c + (a + b - c)/(a + b) + 4.0$;
   float abc (float a, float b, float c)
   {
   return (a + b + b*c + (a+b-c)/(a+b) + 4.0);
   }

The problem instance is characterized by the specific values of a, b, and c. Making
the assumption that one word is adequate to store the values of each of a, b, and c,
and in the result, we see that the space needed by *abc* is independent of the instance
characteristics. Consequently, **Sp(instance characteristics)** = 0.

3. Iterative function for sum
   float Sum(float a[], int n)
   {
   float s = 0.0;
   for (int i=1; i<=n; i++)
   s+=a[i];
   return s;
   }

The problem instances are characterized by n, the number of elements to be summed. The space needed by n is one word, since it is of type *integer*. The space needed by a is the space needed by variables of type array of floats. This is at least n words, since a must be large enough to hold the n elements to be summed. So, we obtain

$$S_{sum}(n) \geq (n+3) \ \ (n \text{ for } a[\ ], \text{ one each for } n, i, \text{ and } s).$$

Time Complexity

The time complexity of an algorithm is the amount of computer time it needs to run to completion.

Reasons for the Time Complexity of a Program

- Some computer systems require the user to provide an upper limit on the amount of time the program will run. Once this upper limit is reached, the program is aborted.
- The program might need to provide a satisfactory real-time response.
- If there are alternative ways to solve a problem, then the decision on which to use will be based primarily on the expected performance difference among these solutions.

The time $T(P)$ taken by a program P is the sum of the compile time and the run (or execution) time. The compile time does not depend on the instance characteristics. A compiled program will run several times without recompilation. The run time is denoted by t_p (instance characteristics).

Because many of the factors t_p depends on are not known at the time a program is conceived, it is reasonable to attempt only to estimate t_p. If we knew the characteristics of the compiler to be used, we could proceed to determine the number of additions, subtractions, multiplications, divisions, compares, loads, stores, and so on that would be made by the code for P. So, we could obtain an expression for $t_p(n)$ of the form

$$t_p(n) = C_a ADD(n) + C_s SUB(n) + C_m MUL(n) + C_d DIV(n) + \ldots \ldots,$$

where n denotes the instance characteristics.

C_a, C_s, C_m, C_d, and so on, respectively, denote the time needed for addition, subtraction, multiplication, division, and so on, and ADD, SUB, MUL, DIV, and so on are functions whose values are the numbers of additions, subtractions, multiplications, divisions, and so on that are performed when the code for P is used on an instance with characteristic n.

In a multi-user system, the execution time depends on such factors as system load, the number of other programs running on the computer at the time program P is run, the characteristics of these other programs, and so on. To overcome the disadvantage of this method, we can go one step further and count only the number of 'program steps', where the time required by each step is relatively independent of the instance characteristics.

A *program step* is defined as a syntactically or semantically meaningful seg-
ment of a program for which the execution time is independent of the instance
characteristics.

The program statements are classified into three types depending on the task to be
performed, and the number of steps required by the unique type is given as follows:

a. Comments—zero step
b. Assignment statement—one step
c. Iterative statement—finite number of steps (for, while, repeat–until)

The number of steps needed by a program to solve a particular problem instance
is done using two different types of methods.

Method 1
A new global variable count is assigned with an initial value of zero. Next, introduce into the program statements to increment the count by the appropriate amount.
Therefore, each time a statement in the original program or function is executed, the
count is incremented by the step count of that statement.

Method 2
To determine the step count of an algorithm a table is built in which we list the total
number of steps contributed by each statement. The final total step count is obtained
by consecutive three steps:

- Calculate the number of *steps per execution (s/e)* of the statement.
- Determine the total number of times each statement is executed (*i.e.,*
  frequency).
- Multiply *(s/e)* and *frequency* to find the total steps of each statement and
  add the total steps of each statement to obtain a *final step count (i.e., total)*.

Example for Method 1
Introduce a variable count in the algorithm sum computes a[i] iteratively, where the
a[i]'s are real numbers.

```
Algorithm sum (a, n)
{
        s:= 0.0;
        count = count + 1; //count is global; it is initially zero.
        for i: = 1 to n do
        {
                count: = count + 1; // For for loop
                s: = s+ a[i];
                count: = count + 1; //For assignment
        }
        count: = count +1; //For last time of for
```

```
        count: = count +1; // For the return;
        return s;
}
```

Step Count

The change in the value of count by the time this program terminates is the number of steps executed by the algorithm sum.

1. Count in the for loop—2n steps
2. For assigning s value to zero—1 step
3. For last time for execution—1 step
4. For return statement—1 step
5. That is, each invocation of sum executes a total of 2n + 3 steps.

Example for Method 2

Find the total step count of the summation of n numbers using the tabular method.

Statement	s/e	Frequency	Total steps
1. Algorithm sum(a,n)	0	–	0
2. {	0		0
3. s=0.0;	1	1	1
4. for i=1 to n do	1	n + 1	n + 1
5. s:=s+a[i];	1	n	n
6. return s;	1	1	1
7. }	0	–	0
Total			2n + 3

1.2 TIME–SPACE TRADE-OFF

Time and space complexity can be reduced only to certain levels, as later on a reduction of time increases the space and vice versa; this is known as time–space trade-off.

Consider the following example shown in Figure 1.5, where we have an array of *n* numbers that are arranged in ascending order. Our task is to get the output as an array that contains these numbers in descending order.

- The first method is by taking two arrays, one for the input and the other for the output.
- Now, read the elements of the first array in reverse linear order and place them in the second array linearly from the beginning.

The code for such an operation is as follows:

```
int ary1 [n];
int ary2[n];
```

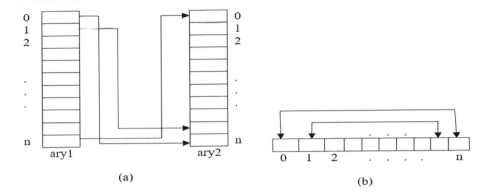

FIGURE 1.5 (a) Reversing array elements using two arrays. (b) Reversing array elements using the swap method.

```
for(int i=0;i<n;i++)
ary2[i]=ary1[(n–1)-i];
```

- Another approach is to just swap the first and last elements, then swap the next two immediate elements as one from each end, and so on.
- This process is repeated until all the elements of the array get swapped.
- Here, no extra array is used. The input array alone gets modified into the output array.

The code for such operation is as follows:

```
int ary1[n];
int k = floor(n/2);
for(int i=0;i<k;i++)
swap(&ary1[i],&ary1[(n–1)-i]) ;
```

where the swap function is

```
swap(int *a, int *b)
{
int temp = *a;
*a=*b;
*b=temp;
}
```

In the first method, an extra array of size n, if n is the size of the input array, is used. The output is obtained by simply assigning values of *ary1* into *ary2*, in reverse order. Therefore, the total space required for the first algorithm is $2n$.

Let us now state the time complexity based on the number of assignment statements.

- Inside the *for* loop, there are *n* assignment statements, and hence, the time complexity of this algorithm is *n* units of time.
- In the second method, only one array of size *n*, and a temporary variable *temp*, inside the swap function, is used.

So, the total space needed by the algorithm is *n* + 1.

- For each swapping, three assignments are required. But the number of times swapping is to be done in at most half of the time of the size of the array.

This leads to the time complexity of *3n/2*.

- In both methods, any attempt to reduce space leads to an increase in the time taken by the algorithm and vice versa.
- The first method increases space with less time, but the second approach reduced space considerably but with an increase in time. This is an example of the time–space trade-off.

1.3   ASYMPTOTIC NOTATIONS

Asymptotic notations are methods used to estimate and represent the efficiency of an algorithm using simple formula. While analyzing the algorithms, three standard notations are commonly used. They are O, Ω, and θ , which are, respectively, named as big oh, big omega, and big theta.

The asymptotic notations are

- **Big-Oh Notation (O),**
- **Big-Omega Notation (Ω),**
- **Big-Theta Notation (θ),**
- **Little-Oh Notation (o),**
- **Little-Omega Notation (ω).**

Let us see the detailed definitions of these notations now.

Big-oh notation (O): This notation is used to define the worst-case running time of an algorithm and is concerned with very large values of *n*.

DEFINITION

Let f and g be two functions defined from a set of natural numbers to a set of nonnegative real numbers. That is f, g: $N \rightarrow R \geq 0$. It is said that the function $f(n) = O(g(n))$ (read as 'f of n is big oh of g of n') if there exists two positive constants $C \in R$ and $n_0 \subset N$ such that $f(n) \leq C g(n)$ for all n, $n \geq n_0$. That is, $f(n)$ grows no faster than $g(n)$; $g(n)$ is the upper bound. The big-oh notation provides an upper bound for the function f. The definition is illustrated in Figure 1.6, where, for the sake of visual clarity, *n* is extended to be a real number.

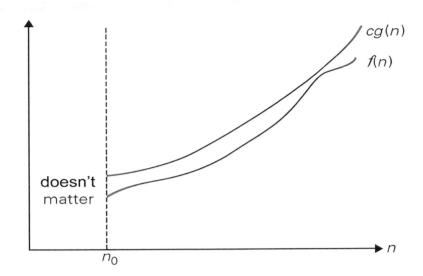

FIGURE 1.6 Big-oh notation: $f(n) \in O(g(n))$

Example 1

Consider $f(n) = 3n + 2$,
where n is at least 2, $3n + 2 < 3n + n$.
$3n + 2 \leq 4n$
So $f(n) = O(n)$,
that is, $f(n) \leq Cg(n)$,
$3n + 2 \leq 4n$, where $C = 4$ and $n_0 = 2$.

Example 2

Consider $f(n) = 100n + 6$.
$100n + 6 < 100n + n$ for all $n \geq 6$.
$100n + 6 \leq 101n$
So $f(n) = O(n)$;
that is, $f(n) \leq Cg(n)$,
$101n + 6 \leq 101n$, where $C = 101$ and $n_0 = 6$.

Example 3

Consider $f(n) = 10n^2 + 4n + 2$,
where $n \geq 2$, $f(n) \leq 10n^2 + 5n$.
$10n^2 + 4n + 2 \leq 10n^2 + 5n$
For $n \geq n_0 = 5$,
$f(n) \leq 10n^2 + n^2$
$10n^2 + 5n \leq 11n^2$
$C = 11$ and $n \geq 5$.

Big-omega notation (Ω): This notation is used to describe the best-case running time of algorithms and is concerned with very large values of n.

DEFINITION

Let f and g be two functions defined from a set of natural numbers to a set of nonnegative real numbers. That is, f, g: $N \rightarrow R \geq 0$. It is said that the function $f(n) = \Omega(g(n))$ (read as 'f of n is omega of g of n'), if there exists two positive constants $C \in R$ and $n_0 \in N$ such that $f(n) \geq C\, g(n)$ for all n, $n \geq n_0$. The definition is illustrated in Figure 1.7. The Omega notation provides the lower bound for the function f.

Example 1

    $f(n) = 3n + 2$ for all n, $3n + 2 > 3n$.
    So $f(n) = \Omega(n)$.

Example 2

    $f(n) = 10n^2 + 4n + 2$
    for all n, $10n^2 + 4n + 2 > 10n^2$.
    So $f(n) = \Omega(n^2)$.
    Also, $3n + 2 > 1$, so $f(n) = \Omega(1)$.
    $10n^2 + 4n + 2 > n$. So $f(n) = \Omega(n)$.

Big-theta notation (θ): This notation is used to define the exact complexity in which the function f(n) is bound both above and below by another function g(n).

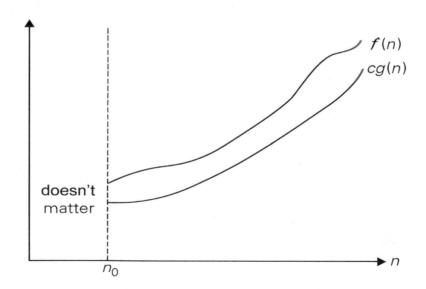

FIGURE 1.7   Big-omega notation: $f(n) \in \Omega(g(n))$

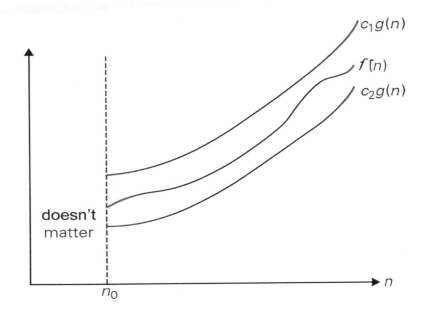

FIGURE 1.8 Big-theta notation: $f(n) \in \theta(g(n))$

DEFINITION

Let f and g be two functions defined from a set of natural numbers to a set of non-negative real numbers. That is, f, g: $N \rightarrow R \geq 0$. It is said that the function $f(n) = \theta$ (g(n)) (read as 'f of n is theta of g of n'), if there exists three positive constants C_1, $C_2 \in R$ and $n_0 \in N$ such that $C_1 g(n) \leq f(n) \leq C_2\ g(n)$ for all n, $n \geq n_0$. The definition is illustrated in Figure 1.8.

Example

Consider f(n) = 3n + 2.

$$4n \geq 3n + 2 \geq 3n$$
$$3n \leq 3n + 2 \leq 4n$$
$$C_1\ g(n) \leq f(n) \leq C_2\ g(n)$$
$$C_1 = 3\ C_2 = 4 \text{ Therefore, } 3n + 2 = \theta(n).$$

Little-oh notation (o): This notation is used to describe the worst-case analysis of algorithms and is concerned with small values of n.

The function f(n) = o(g(n)) (read as 'f of n is little oh of g of n') iff

$$\lim_{n \to \infty} \frac{f(n)}{g(n)} = 0.$$

Little-omega notation (ω): This notation is used to describe the best-case analysis of algorithms and is concerned with small values of n.

The function $f(n) = \omega(g(n))$ (read as 'f of n is little omega of g of n') iff

$$\lim_{n\to\infty} \frac{g(n)}{f(n)} = 0.$$

1.4   PROPERTIES OF BIG-OH NOTATIONS

STATEMENT 1

$$O(f(n)) + O(g(n)) = O(\max\{f(n),g(n)\})$$

Proof

Let $f(n) \le g(n)$.
L.H.S $= C_1 g(n) + C_2 f(n)$
$\le C_1 g(n) + C_2 g(n)$
$= (C_1 + C_2)g(n)$
$= O(g(n))$
$= O(\max\{f(n),g(n)\})$

STATEMENT 2

$f(n) = O(g(n))$ and $g(n) \le h(n)$ implies $f(n) = O(h(n))$.

STATEMENT 3

Any function can be said as an order of itself. That is, $f(n) = O(f(n))$.

Proof

Proof of this property trivially follows from the fact that $f(n) \le 1 \times f(n)$.

STATEMENT 4

Any constant value is equivalent to $O(1)$. That is $C = O(1)$, where C is a constant. The asymptotic notations are defined only in terms of the size of the input.

For example, the size of the input for sorting n numbers is n. So, the constants are not directly applied to the stated facts to obtain the results. For instance, suppose we have the summation $1^2 + 2^2 + \ldots\ldots\ldots\ldots + n^2$. By using the property explained in statement 1 directly, we get

$$1^2 + 2^2 + \ldots\ldots\ldots\ldots + n^2 = O\,(\max\{1^2, 2^2, \ldots\ldots, n^2\})$$
$$= O(n^2), \text{ which is wrong.}$$

Actually, the asymptotic value of the sum is $O(n^3)$ and is obtained as follows:

$$1^2 + 2^2 + \ldots\ldots\ldots\ldots + n^2 = n(n + 1)(2n + 1)/6$$
$$= (2n^3 + 3n^2 + n)/6$$
$$= O(n^3) + O(n^2) + O(n)$$
$$= O(\max\{n^3, n^2, n\})$$
$$= O(n^3)$$

STATEMENT 5

$$\text{If } \lim_{n\to\infty}\{f(n)/g(n)\} \in R > 0 \text{ then } f(n) \in \theta\big(g(n)\big).$$

STATEMENT 6

$$\text{If } \lim_{n\to\infty}\left\{\frac{f(n)}{g(n)}\right\} = 0 \text{ then } f(n) \in O\big(g(n)\big) \text{ but } f(n) \notin \theta\big(g(n)\big).$$

$$\text{That is } O\big(f(n)\big) \subset O\big(g(n)\big),$$

which also implies $f(n) \in O\big(g(n)\big)$ and $g(n) \notin O\big(f(n)\big)$.

STATEMENT 7

$$\text{If } \lim_{n\to\infty}\{f(n)/g(n)\} = +\infty => f(n) \in \Omega\big(g(n)\big) \text{ but } f(n) \notin \theta\big(g(n)\big).$$

$$\text{That is, } O\big(f(n)\big) \supset O(g(n)).$$

Sometimes, the L'Hospital's rule is useful in obtaining the limit value. The rule states that

$$\lim_{n\to\infty}\frac{f(n)}{g(n)} = \lim_{n\to\infty}\frac{f'(n)}{g'(n)},$$

where f' and g' are the first derivatives of f and g, respectively.

PROBLEM

Prove that $\log n \in O(\sqrt{n})$ but $\sqrt{n} \notin O(\log n)$.

Proof

$$\lim_{n\to\infty}\frac{\log n}{\sqrt{n}} = \lim_{n\to\infty}\frac{\frac{1}{n}}{\frac{1}{2}n^{\frac{-1}{2}}} \qquad\qquad \text{[By L-Hospital's rule]}$$

$$= \lim_{n\to\infty}\frac{2\sqrt{n}}{n}$$

$$= \lim_{n \to \infty} \frac{2}{\sqrt{n}}$$

$$= 0$$

Therefore, by statement 6, $\log n \in O(\sqrt{n})$ but $\sqrt{n} \notin O(\log n)$.

- The asymptotic values of certain functions can be easily derived by imposing certain conditions.
- For example, if the function f(n) is recursively defined as

$$f(n) = sf(n/2) + n, f(1) = 1.$$

Suppose n is the power of 2. That is, $n = 2^k$ for some k.

$$\begin{aligned} f(2^k) &= 2f(2^{k-1}) + 2^k \\ &= 2^2 f(2^{k-2}) + 2^k + 2^k \\ &= 2^2 f(2^{k-2}) + 2(2^k) \\ &\qquad \cdots \cdots \cdots \cdots \\ &\qquad \cdots \cdots \cdots \cdots \\ &= 2^k f(1) + k2^k, \end{aligned}$$

which is equal to $n + n \log n$, and so, O (n log n), where $n = 2^k$.

1.5   CONDITIONAL ASYMPTOTIC NOTATIONS

Many algorithms are easier to analyze if we impose conditions on them initially. Imposing such conditions, when we specify an asymptotic value, is called conditional asymptotic notation.

CONDITIONAL BIG-OH NOTATION

Let f,g: N → R ≥ 0. It is said that f(n) = O(g(n)|A(n)), read as g(n) when A(n), if there exist two positive constants $C \in R$ and $n_0 \in N$ such that $A(n) \Rightarrow [f(n) \leq Cg(n)]$ for all $n \geq n_0$.

CONDITIONAL BIG-OMEGA NOTATION

Let f,g: N → R ≥ 0. It is said that f(n) = Ω(g(n)|A(n)), read as g(n) when A(n), if there exist two positive constants $C \in R$ and $n_0 \in N$ such that $A(n) \Rightarrow [f(n) \geq Cg(n)]$ for all $n \geq n_0$.

CONDITIONAL BIG-THETA NOTATION

Let f,g: N → R ≥ 0. It is said that f(n) = θ (g(n)|A(n)), read as g(n) when A(n), if there exists two positive constants C_1, $C_2 \in R$ and $n_0 \in N$ such that $A(n) \Rightarrow [C_1 g(n) \leq f(n) \leq$

C_2 g(n)] for all n ≥ n_0. We know that f(n) = 2f(n/2) + n = O(n log n| n is a power of 2), which is the conditional asymptotic value. That is, f(n) = O(n log n) $\forall n \geq n_0$ for some $n_0 \in N$.

A function f: N → R ≥ 0 is **eventually nondecreasing** if $\exists n_0 \in N$ such that f(n) ≤ f(n + 1), $\forall n \geq n_0$.

THEOREM

Let p ≥ 2 be an integer. Let f, g: N → R ≥ 0. Also, f be an eventually nondecreasing function and g be a p-smooth function. If f(n) ∈ O(g(n)| n is a power of p), then f(n) ∈ O(g(n)).

Proof

Apply the theorem for proving f(n) = O(n log n), $\forall n$.

To prove this, we have to prove that f(n) is eventually nondecreasing and n log n is 2-smooth.

Claim 1: f(n) is eventually nondecreasing.

Using mathematical induction, the proof is as follows

$$f(1) = 1 \leq 2(1) + 2 = f(2).$$

Assume for all m < n, f(m) ≤ f(m + 1).

In particular, f(n/2) ≤ f((n + 1)/2).

Now,

$$f(n) = 2 \, f(n/2) + n$$
$$\leq 2f((n + 1)/2) + (n + 1)$$
$$= f(n + 1).$$

Therefore, f is eventually nondecreasing.

Claim 2: n log n is 2-smooth.

$$2n \log 2n = 2n(\log 2 + \log n)$$
$$= O(n \log n), \text{ which implies n log n is 2-smooth.}$$

Therefore, f(n) = O(n log n).

The preceding function f(n) is simple in obtaining the asymptotic value by recursively applying the function.

1.6 RECURRENCE EQUATIONS

A recurrence is an equation or an inequality that describes a function in terms of its values on the smallest inputs.

DERIVING RECURRENCE RELATIONS

To derive a recurrence relation,

- figure out what 'n', the problem size, is.
- see what value of n is used as the base of the recursion. It will usually be a single value but may be multiple values suppose it is n_0.
- figure out what $T(n_0)$ is, you can usually use 'constant C', but sometimes a specific number will be needed.
- the general T(n) is usually a sum of various choices of T(m) (for the recursive calls) plus the sum of the other work done. Usually, the recursive calls will be solving subproblems of the same size f(n), giving a term 'a.T(f(n))' in the recurrence relation.

$$T(n) = \begin{cases} C & if \, n = n_0 \\ a.T(f(n)) + g(n) & otherwise, \end{cases}$$

where C = running time for base,

n_0 = base of recurrence,

a = the number of times a recursive call is made,

f = the size of the problem solved by a recursive call, and

g(n) = all other processing not counting recursive calls.

Recurrence equations can be classified into homogeneous and inhomogeneous.

In algebra, there are easy ways to solve both homogeneous and inhomogeneous equations.

Suppose **T(n)** is the time complexity of our algorithm for the size of the input **n**. Assume that **T(n)** is recursively defined as

$$T(n) = b_1 T(n-1) + b_2 T(n-2) + \ldots + b_k T(n-k)$$

$$a_0 T(n) + a_1 T(n-1) + \ldots + a_k T(n-k) = 0. \tag{1}$$

The constants, which are b_is, are now converted to a_is for simplicity.

Let us denote T(i) as x^i. Hence, the equation becomes

$$a_0 X^n + a_1 X^{n-1} + \ldots + a_k X^{n-k} = 0, \tag{2}$$

which is a homogeneous recurrence equation.

One of the trivial solutions for Equation 2 is x = 0.

After removing common x terms from Equation 2, we get

$$a_0 X^k + a_1 X^{k-1} + \ldots + a_k = 0.$$

The preceding equation is known as a characteristic equation and can have k roots.

Let the roots be $r_1, r_2, \ldots \ldots, r_k$. The roots may not be the same.

Case 1: Suppose all the roots are distinct.
Then, the general solution is

$$T(n) = \sum_{i=1}^{k} c_i r_i^n, \text{ where } c_i\text{'s are some constants.}$$

For example, suppose the characteristic equation is $x^2 - 5x + 6 = 0$. Then

$$
\begin{aligned}
& x^2 - 5x + 6 = 0 \\
\Rightarrow \quad & (x - 3)(x - 2) = 0 \\
\Rightarrow \quad & \text{the roots are 3 and 2.}
\end{aligned}
$$

Therefore, the general solution is

$$T(n) = c_1 3^n + c_2 2^n.$$

Case 2: Suppose some of the p roots are equal and the remaining are distinct.
Let us assume that the first p roots are equal to r_1. Then, the general solution can be stated as

$$T(n) = \sum_{l=1}^{p} c_i n^{i-1} r_1^n + \sum_{i=p+1}^{k} c_i r_i^n.$$

For example, suppose the characteristic equation is $(x - 2)^3 (x - 3) = 0$.
Then the roots of this equation are 2, 2, 2, and 3. Therefore, the general solution is

$$T(n) = c_1 2^n + c_2 n 2^n + c_3 n^2 2^n + c_4 3^n.$$

Case 2 can be extended to cases like p roots are equal to r_1, q roots are equal to r_2, and so on.

INHOMOGENEOUS RECURRENCE EQUATION

The general form is as follows

$$a_0 t_n + a_1 t_{n-1} + \ldots\ldots\ldots + a_k\, t_{n-k} = b_1^n P_1(n) + b_2^n P_2(n) + \ldots,$$

where a_i's and b_i's are constants. Each $P_i(n)$ is a polynomial in n of degree d_i.
Here the characteristic equation tends to be

$$
\begin{aligned}
& (a_0 x^k + a_1 x^{k-1} + \ldots\ldots\ldots + a_k)(x - b_1)^{d1+1}(x - b_2)^{d2+1} \ldots = 0 \\
\Rightarrow\ & a_0 x^k + a_1 x^{k-1} + \ldots\ldots\ldots + a_k = 0 \\
& (x - b_1)^{d1+1} = 0 \\
& (x - b_2)^{d2+1} = 0 \\
& \ldots\ldots\ldots\ldots\ldots
\end{aligned}
$$

These are a set of homogeneous equations, and hence, the general solution can be obtained as before.

Let the general solutions for the earlier equations be $S_1, S_2, \ldots$ Then the general solution of the inhomogeneous equation is given by

$$T(n) = S_1 + S_2 + \ldots$$

Example

The recurrence equation of the complexity of merge-sort algorithm

$$f(n) = 2f(n/2) + n, f(1) = 1 \tag{1}$$

Let $n = 2^k$.
For simplicity, we say $f(2^k) = t_k$. Therefore,

$$\text{Equation 1} \Rightarrow t_k - 2t_{k-1} = 2^k \tag{2}$$

Equation 2 is an inhomogeneous equation.
The characteristic equation of the recurrence Equation 2 is

$$(x - 2)(x - 2) = 0$$
$$\Rightarrow (x - 2)^2 = 0$$

The roots are 2 and 2. Now the general solution is

$$t_k = c_1 2^k + c_2 k 2^k$$
$$\Rightarrow f(n) = c_1 n + c_2 n \log n \tag{3}$$
$$\text{Given that } f(1) = 1 \Rightarrow 1 = c_1,$$
$$\text{similarly, } f(2) = 2f(1) + 2 = 2 + 2 = 4.$$

From Equation 3,
$$4 = c_1 2 + c_2 2$$
$$\Rightarrow 4 = 2 + c_2 2$$
$$\Rightarrow c_2 = 1.$$

Therefore, Equation 3 becomes

$$f(n) = n + n \log n$$
$$= O(n \log n| n \text{ is a power of 2}).$$

As we have already proved that $n \log n$ is 2-smooth and that (n) is eventually nondecreasing.

So we say $f(n) = O(n \log n)$.

1.7   SOLVING RECURRENCE EQUATIONS

There are three methods of solving recurrence relations:

- Substitution
- Iteration
- Master

SUBSTITUTION METHOD

In the substitution method, make intelligent guesswork about the solution of a recurrence and then prove by induction that the guess is correct.

Given a recurrence relation $T(n)$,

- substitute a few times until you see a pattern.
- write a formula in terms of n and the number of substitutions i.
- choose I so that all references to $T()$ become references to the base case.
- solve the resulting summation.

Example
Multiplication for all $n > 1$

$$T(n) = T(n-1) + d$$

Therefore, for a large enough n,

$$T(n) = T(n-1) + d$$
$$T(n-1) = T(n-2) + d$$
$$.$$
$$.$$
$$.$$
$$T(2) = T(1) + d$$
$$T(1) = c.$$

Repeated substitution

$$T(n) = T(n-1) + d$$
$$= (T(n-2) + d) + d$$
$$= T(n-2) + 2d$$
$$= (T(n-3) + d) + 2d$$
$$= T(n-3) + 3d$$

There is a pattern developing. It looks like after i substitutions,

$$T(n) = T(n-i) + id$$

Now choose $i = n - 1$, then

$$
\begin{aligned}
T(n) &= T(n - (n - 1)) + (n - 1)d \\
&= T(n - n + 1) + (n - 1)d \\
&= T(1) + (n - 1)d \\
&= c + nd - d \\
T(n) &= dn + c - d
\end{aligned}
$$

Iteration Method

In the iteration method, the basic idea is to

- expand the recurrence,
- express it as a summation of terms dependent only on **n** and the initial conditions, and
- evaluate the summation.

Example: Consider the recurrence $T(n) = T(n - 1) + 1$, and $T(1) = \theta(1)$.

Solution

$$
\begin{aligned}
T(n) &= T(n - 1) + 1 \\
T(n - 1) &= T(n - 2) + 1 \\
T(n) &= (T(n - 2) + 1) + 1 \\
&= T(n - 2) + 2 \\
T(n - 2) &= T(n - 3) + 1 \\
\text{Thus, } T(n) &= (T(n - 3) + 1) + 2 \\
&= T(n - 3) + 3 \\
T(n) &= T(n - k) + k \\
k &= n - 1 \\
T(n - k) &= T(n - (n - 1)) \\
&= T(n - n + 1) \\
&= T(1) \\
T(n) &= T(1) + k \\
&= T(1) + k \\
T(n) &= \theta(n)
\end{aligned}
$$

Master Method

The master method is used for solving the following type of recurrence:

$$
T(n) = aT(n/b) + f(n), \text{ with } a \geq 1 \text{ and } b > 1.
$$

In the preceding recurrence, the problem is divided into a subproblem each of size almost (n/b). The subproblems are solved recursively each in $T(n/b)$ time. The cost of splitting the problem or combining the solutions of subproblems is given by function $f(n)$. It should be noted that the number of leaves in the recursion tree is n^E with $E = (\log a/b \log b)$.

Master Theorem

Let T(n) be defined on the nonnegative integers by the recurrence.

$$T(n) = aT(n/b) + f(n),$$

where a ≥ 1 and b > 1 are constants, f(n) is a function, and n/b can be interpreted as [n/b]. Then T(n) can be bound asymptotically as follows:

Case 1: If $f(n) \in O\left(n^{E-e}\right)$ for some e > 0, then $T(n) \in \theta(n^E)$.

Case 2: If $f(n) \in \theta\left(n^E\right)$, then $T(n) \in \theta\left(f(n).logn\right)$.

Case 3: If $f(n) \in \Omega\left(n^{E+e}\right)$ for some e > 0 and $f(n) \in O\left(n^E + \delta\right)$ for some δ ≥ e, then $T(n) \in \theta\left[\left(f(n)\right)\right]$.

Example

Consider the following recurrence T(n) = 4T(n/2) + n.
  Using the master method,

$$a = 4, b = 2, \text{ and } f(n) = n$$

$$n^F - n^{(\log\ a/\log b)}, \text{ where } E = \frac{\log a}{\log b}.$$

$$= n\ \log_b a$$
$$= n\ \log 24$$
$$= n^2$$

Since f(n) = n ∈ $O(n^{2-e})$,
  thus, T(n) ∈ $\theta(n^2)$.

Problem 1: Solve the recurrence equation T(n) − 2T(n − 1) = 3^n subject to T(0) = 0.

Proof

  The characteristic equation is (x − 2)(x − 3) = 0.
  Therefore, the roots are 2 and 3.

Now the general solution is T(n) = $c_1 2^n + c_2 3^n$.
  Since T(0) = 0, T(1) = 3. Thus, from the general solution, we get c_1 = −3 and c_2 = 3.
  So, T(n) = −3 × 2^n + 3 × 3^n = $O(3^n)$

Problem 2: Solve the recurrence equation T(n) − 2T(n − 1) = 1 subject to T(0) = 0.

Proof

  The characteristic equation is (x − 2)(x − 1) = 0.
  Therefore the roots are 2 and 1.

Now the general solution is $T(n) = c_1 1^n + c_2 2^n$.

Since $T(0) = 0$, $T(1) = 1$. Thus, from the general solution, we get $c_1 = -1$ and $c_2 = 1$.
So, $T(n) = 2^n - 1 = \theta(2^n)$.

Problem 3: Solve the recurrence equation $T(n) = 2T(n-1) + n2^n + n^2$.

Proof

The characteristic equation is $(x-2)(x-2)^2(x-1)^3 = 0$.
$(x-2)^3(x-1)^3 = 0$

Therefore the roots are 2, 2, 2, 1, 1, and 1
Now the general solution is $T(n) = c_1 2^n + c_2 n2^n + c_3 n^2 2^n + c_4 1^n + c_5 n1^n + c_6 n^2 1^n$.
So $T(n) = O(n^2 2^n)$.

1.8   ANALYSIS OF LINEAR SEARCH

- Algorithms are analyzed to get best-case, worst-case, and average-case asymptotic values.
- Each problem is defined on a certain domain. From the domain, we can derive an instance for the problem.
- In the example of multiplying two integers, any two integers become an instance of the problem.
- So, when we go for analysis, it is necessary that the analyzed value will be satisfiable for all instances of the domain.
  - Let D_n be the domain of a problem, where n be the size of the input.
  - Let $I \in D_n$ be an instance of the problem taken from the domain D_n.
  - $T(I)$ be the computation time of the algorithm for the instance $I \in D_n$.

BEST-CASE ANALYSIS

- The best-case efficiency of an algorithm is its efficiency for the best-case input of size n, which is an input of size n for which the algorithm runs fastest among all possible inputs of that size.
- Determine the kind of inputs for which the count $c(n)$ will be the smallest among all the possible inputs of size n.
- This gives the minimum computed time of the algorithm with respect to all instances from the respective domain.
  Mathematically, it can be stated as
    $B(n) = \min\{T(I) \mid I \in D_n\}$.
- If the best-case efficiency of an algorithm is unsatisfactory, we can immediately discard it without further analysis.

WORST-CASE ANALYSIS

- The worst-case efficiency of an algorithm is its efficiency for the worst-case input of size n, which is an input of size n for which the algorithm runs the longest among all possible inputs of that size.

- Analyze the algorithm to see what kind of inputs yield the largest value of basic operation's count c(n) among all the possible inputs of size n and then compute this worst-case value $C_{worst}(n)$.
- This gives the maximum computation time of the algorithm with respect to all instances from the respective domain.
- Mathematically, it can be stated as
  - $W(n) = \max\{T(I) \mid I \in \mathbf{D_n}\}$.

AVERAGE-CASE ANALYSIS

- It is neither the worst-case analysis nor its best-case counterpart yields the necessary information about an algorithm's behavior on a 'typical' or 'random' inputs.
- To analyze the algorithm's average-case efficiency, make some assumptions about possible inputs of size **n**.
- Mathematically, the average case value can be stated as

$$A(n) = \sum_{I \in D_n} P(I)T(I).$$

- P(I) is the average probability with respect to instance I.

Example

- Take an array of some elements as shown in Figure 1.9. The array is to be searched for the existence of an element x.
- If the element x is in the array, the corresponding location of the array should be returned as an output.
- The element may or may not be in the array. If the element is in the array, it may be in any location, that is, in the first location, the second location, anywhere in the middle, or at the end of the array.
- To find the location of the element *x*, we use the following algorithm, which is said to be the ***linear search*** algorithm.

```
int linearsearch(const char A[ ], const unsigned int size, char ch)
{
    for(int i=0;i<size; i++)
    {
            if (A[i]== ch)
            return(i);
    }
    return (−1);
}
```

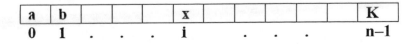

FIGURE 1.9   A sample array of elements

TABLE 1.1

Number of Comparisons to Find the Element of the Array

Location of the element	Number of comparisons required
0	1
1	2
2	3
.	.
.	.
.	.
n − 1	n
Not in the array	n

- The searching process done here is sequential as one location after the other is searched beginning from the first location to the end of the array, and hence, it is named a linear search.
- The program terminates as soon as it finds the element or fails to find the element in the array after searching the whole array.
- The return value for the successful search is the location of the element x, and for an unsuccessful search, it is **−1**.
- In this program, the comparison is made between the array element and the element that is to be searched for.

Table 1.1 provides information on the number of comparisons required to find the element present in various locations of the array.

Best Case

The best case arises when the searching element appears in the first location (0).

$$B(n) = \min\{1, 2, \ldots \ldots, n\}$$
$$= 1$$
$$= O(1)$$

Worst Case

In the worst case, the element could either be in the last location or could not be in the array.

$$W(n) = \max\{1, 2, \ldots, n\}$$
$$= n$$
$$= O(n)$$

Average Case

Let k be the probability of x being in the array. Therefore, the average probability is

$$P(I_i) = k/n \text{ for } 0 \le i \le n - 1$$
$$P(I_n) = 1 - k$$

The preceding equation is the probability of x not being in the array.

$$A(n) = \sum_{i=0}^{n} P(I_i)T(I_i)$$

$$= (k/n)\sum_{i=0}^{n-1}(i+1) + (1-k)n$$

$$= \frac{k}{n}\frac{n(n+1)}{2} + (1-k)n$$

$$= \frac{k(n+1)}{2} + (1-k)n$$

Suppose x is in the array, then k = 1. Therefore,

$$A(n) = \frac{n+1}{2} = O(n).$$

In case of x being in the array or not, k = 1/2

$$\Rightarrow A(n) = \frac{(n+1)}{4} + \frac{n}{2} = \left(\frac{3}{4}\right)n + 1 = O(n)$$

Table 1.2 concludes the preceding discussion.

Mostly, we write programs directly than writing pseudo-code algorithms. Sometimes analyzing programs will be quite simple due to their control structures. Some of these control structures and the asymptotic time it takes to run are given in the following sections.

LOOP CONTROLS

The loop control statements in C++ can be classified into two categories, exit-control and entry-control. The name is due to the place where the condition is verified. As shown in Figure 1.10, the *do-while* loop comes under exit control as the condition is verified at the end of the loop in each iteration. Similarly, the *while* loop and *for* loop comes under the category of entry control as the condition is verified at the beginning of each iteration.

TABLE 1.2

Computation Time for Linear Searching

Algorithm	Best case	Worst case	Average case
Linear Search	O(1)	O(n)	O(n)

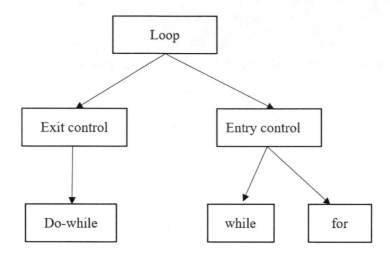

FIGURE 1.10   Loop controls

Let us see the asymptotic running time of the following programming statements.

```
int count = 0;
for (int i – 0;i<n;i++)
count++;
```

Here, the rule states that *the running time of the* **for** *loop* = **max** *{running time of the statements inside the* **for** *loop} × number of iterations.*

Therefore, the *'for'* loop in the preceding code takes O(n) time. This set of statements can be written using either *do-while* or *while* loop, which also takes the same complexity:

```
int count =0;
do
{
count++;
}while (count==n);
(or)
int count = 0;
while (count != n)
count++;
```

Suppose one *for* loop lies inside another *for* loop as in the following example:

```
int count = 0;
for (int i = 0;i<n;i++)
        for(int j = 0;j < m; j++)
                count++;
```

Here, the maximum running time of the statements inside the outer loop is O(m). Therefore, the total time needed to run this code is O(nm).

Another rule says that in the case of a sequence of instructions, we should add the computation time for each statement in that sequence.

Example

```
int count = 0;
for (int i = 0;i<n;i++)
        count++;
for(int j = 0;j < m; j++)
        count--;
```

The first *for* loop takes O(n) time.
The second *for* loop takes O(m) time.
Total time = O(n + m)

$$O(n + m) = O(\max\{m,n\})$$

For the *if–then–else* statement,

if (condition)
statementl;
else
statement2;

- Running time = *Time(testing conditions+* max *{Time(statementl), Time(statement2)).*
- If there are recursive calls in the code, each call must be counted for the purpose of analysis.

For instance, suppose the following code:

```
unsigned long int Factorial(constant unsigned int n)
{
if (n ≥1)
return 1;
else
return n * Factorial(n-l);
}
```

It takes T(n) time. Then, T(n) = T(n − 1) + 2.
The value 2 accounts for the testing condition and the computation at the second return statement.

Now, we say trivially that $T(n) = O(n)$.

ANALYZING THE EFFICIENCY OF RECURSIVE AND NON-RECURSIVE ALGORITHMS

A General Plan for Analyzing the Efficiency of Recursive Algorithms

1. Decide on a parameter indicating an input's size.
2. Identify the algorithm's basic operation.
3. Check whether the number of times the basic operation is executed can vary on different inputs of the same size. If it can, the worst-case, average-case, and best-case efficiencies must be investigated separately.
4. Set up a recurrence relation with an appropriate initial condition for the number of times the basic operation is executed.
5. Solve the recurrence or at least ascertain the order of growth of its solution.

Example: Tower of Hanoi Problem

Let us apply the general plan to the tower of Hanoi problem. The number of disks **n** is the obvious choice for the input size indicator, and so is moving one disk at the algorithm's basic operation.

The number of moves **m(n)** depends on **n** only and gets the following recurrence equation for it:

$$M(n) = M(n - 1) + 1 + M(n - 1) \text{ for } n > 1$$
$$\text{The initial condition } M(1) = 1.$$
$$M(n) = 2M(n - 1) + 1 \text{ for } n > 1$$
$$M(1) = 1$$

Solve this recurrence by substitution method.

$$M(n) = 2M(n - 1) + 1$$
$$M(n - 1) = 2M(n - 2) + 1$$
$$\text{Thus, } M(n) = 2[2M(n - 2) + 1] + 1$$
$$= 2M(n - 2) + 2 + 1$$
$$M(n - 2) = 2M(n - 3) + 1$$
$$\text{Thus, } M(n) = 2^2[2M(n - 3) + 1] + 2 + 1$$
$$= 2^3 M(n - 3) + 2^2 + 2 + 1.$$

After i substitutions, we get
$$M(n) = 2^i M(n - i) + 2^{i-1} + 2^{i-2} + \ldots + 1$$
$$= 2^i M(n - i) + 2^i - 1.$$

If $i = n - 1$,

$$M(n) = 2^{n-1} M(n - (n - 1)) + 2^{n-1} - 1$$
$$= 2^{n-1} M(n - n + 1) + 2^{n-1} - 1$$
$$= 2^{n-1} M(1) + 2^{n-1} - 1$$
$$= 2^{n-1} + 2^{n-1} - 1$$
$$= 2^{n-1}[1 + 1] - 1$$
$$= 2^{n-1}(2) - 1$$
$$M(n) = 2^n - 1.$$

A General Plan for Analyzing Efficiency of Non-Recursive Algorithms

1. Decide on a parameter indicating an input's size.
2. Identify the algorithm's basic operation.
3. Check whether the number of times the basic operation is executed depends only on the size of an input. If it also depends on some additional property, the worst-case, average-case, and, if necessary, best-case efficiencies have to be investigated separately.
4. Set up a sum expressing the number of times the algorithm's basic operation is executed.
5. Using standard formulas and rules of sum manipulation either find a closed-form formula for the count or, at the very least, establish its order of growth.

Example

Consider the matrix multiplication program.

An input's sizes by matrix order n consider multiplication as the algorithm's basic operation.

Let us set up a sum for the total number of multiplications *m(n)* executed by the algorithm.

$$M(n) = \sum_{i=0}^{n-1}\sum_{j=0}^{n-1}\sum_{k=0}^{n-1} 1$$

$$= \sum_{i=0}^{n-1}\sum_{j=0}^{n-1} n$$

$$= \sum_{i=0}^{n-1} n^2$$

$$M(n) = n^3$$

2 Divide and Conquer

2.1 DIVIDE AND CONQUER: GENERAL METHOD

Divide-and-conquer algorithms can provide a small and powerful means to solve a problem; this section is not about how to write such an algorithm but rather how to analyze one. When we count comparisons that occur in loops, we only need to determine how many comparisons there are inside the loop and how many times the loop is executed. This is made more complex when a value of the outer loop influences the number of passes of an inner loop. A technique for designing algorithms that decompose an instance into smaller sub-instances of the same problem, solve the sub-instances independently, and combine the sub-solutions to obtain the solution of the original instance is called divide and conquer.

The divide-and-conquer approach is a top-down approach; that is, the solution to a top-level instance of a problem is obtained by going down and obtaining solutions to smaller instances as shown in Figure 2.1.

Formally, the divide-and-conquer algorithm consists of the following major phases or methodology:

Divide: Break the problem into several subproblems that are similar to the original problem but smaller in size.
Conquer: Solve the subproblems recursively.
Combine: Combine the solutions of the subproblems to create the solution to the original problem.

An analysis can be done using recurrence equations.

General Method

Given a function with **n** inputs, the *divide-and-conquer* strategy splits the input into **k** distinct subsets, $1 < k \leq n$, which yields **k** subproblems. If the subproblems are still relatively large, then the divide-and-conquer strategy can possibly be reapplied. Now smaller and smaller subproblems of the same kind are generated until the subproblems that are small enough to be solved without splitting are produced. These subproblems must be solved, and then a method must be found to combine sub-solutions into a solution of the whole.

```
Type DAndC(P)
{
        if Small(P) return S(P);
        else {
                divide P into smaller instances P₁, P₂ . . . . . . . Pₖ, k≥1;
                Apply DAndC to each of these subproblems;
```

DOI: 10.1201/9781003355403-2

```
        return Combine(DAndC(P₁),DAndC(P₂), . . . ,DAndC(Pₖ));
        }
}
```

DAndC is a function that is initially invoked as DAndC(P), where P is the problem to be solved. Otherwise, problem P is divided into smaller subproblems. These sub-problems $P_1, P_2, \ldots, P_k$ is solved by recursive applications of DAndC.

Small(P) is a Boolean-valued function that determines whether the input size is small enough that the answer can be computed without splitting.

Combine is a function that determines the solution to P using the solutions to the **k** subproblems.

If the size of P is *n* and the sizes of the **k** subproblems are $n_1, n_2, \ldots, n_k$, respectively, then the computing time of DAndCis described by the recurrence relation

$$T(n) = \begin{cases} g(n) & n \ small \\ T(n_1) + T(n_2) + \ldots + T(n_k) + f(n) & otherwise, \end{cases}$$

where **T(n)** is the time for DAndC on any input of size **n**,

g(n) is the time to compute the answer directly for small inputs, and
f(n) is the time for dividing P and combining the solutions to subproblems.

For divide-and-conquer-based algorithms that produce subproblems of the same type as the original problem, then we can apply recursion.

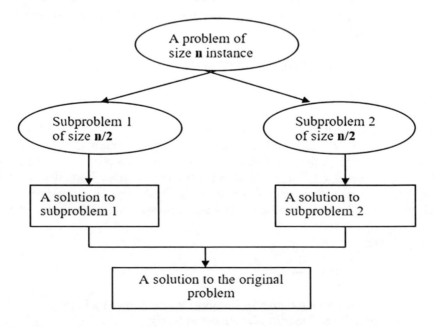

FIGURE 2.1   Divide-and-conquer technique

SOLVING RECURRENCE RELATIONS

The complexity of many divide-and-conquer algorithms is given by recurrences of the form

$$T(n) = \begin{cases} T(1) & n = 1 \\ aT(n/b) + f(n) & n > 1 \end{cases},$$

where **a** and **b** are known constants.

Assume that **T(1)** is known and **n** is a power of **b** (i.e., $n = b^k$).

One of the methods for solving any such recurrence relation is called the ***substitution method***. This method repeatedly makes a substitution for each occurrence of the function **T** in the right-hand side until all such occurrences disappear.

Example: Detecting a Counterfeit Coin

You are given a bag with 16 coins and told that one of these coins may be counterfeit. Furthermore, you are told that counterfeit coins are lighter than genuine ones. Our task is to determine whether the bag contains a counterfeit coin. To aid you in this task, you have a machine that compares the weights of two sets of coins and tells you which set is lighter or whether both sets have the same weight. There are two ways to find the counterfeit coin

One way is to compare the weights of coins 1 and 2. If coin 1 is lighter than coin 2, then coin 1 is counterfeit, and we are done with our task. If coin 2 is lighter than coin 1, then coin 2 is counterfeit. If both coins have the same weight, we compare coins 3 and 4. Again, if one coin is lighter, a counterfeit coin has been detected, and we are done. If not, we compare coins 5 and 6. Proceeding in this way, we can determine whether the bag contains a counterfeit coin by making at most eight weight comparisons. This process also identifies the counterfeit coin.

Another approach is to use the divide-and-conquer methodology. Suppose that our 16-coin instance is considered a large instance.

- In step 1, we divide the original instance into two or more smaller instances. Let us divide our 16-coin instance into two 8-coin instances by arbitrarily selecting 8 coins for the first instance (say **A**) and the remaining 8 coins for the second instance (say **B**).
- In step 2, we need to determine whether **A** or **B** has a counterfeit coin. For this step, we use our machine to compare the weights of the coin sets **A** and **B**. If both sets have the same weight, a counterfeit coin is not present in the 16-coin set. If **A** and **B** have different weights, a counterfeit coin is present, and it is in the lighter set.
- Finally, in step 3 we take the results from step 2 and generate the answer for the original 16-coin instance. For the counterfeit-coin problem, step 3 is easy. The 16-coin instance has a counterfeit coin iff either *A* or *B* has one. So with just one weight comparison, we can complete the task of determining the presence of a counterfeit coin.

Solution

- Now suppose we need to identify the counterfeit coin. The 16-coin instance is a large instance. So it is divided into two 8-coin instances A and B as earlier.
- By comparing the weights of these two instances, we determine whether a counterfeit coin is present. If not, the algorithm terminates. Otherwise, we continue with the sub-instance known to have the counterfeit coin.
- Suppose B is the lighter set. It is divided into two sets of four coins each. Call these sets $B1$ and $B2$. The two sets are compared. One set of coins must be lighter.
- If $B1$ is lighter, the counterfeit coin is in $B1$, and $B1$ is divided into two sets of two coins each. Call these sets $B1a$ and $B1b$. The two sets are compared, and we continue with the lighter set.
- Since the lighter set has only two coins, it is a small instance. Comparing the weights of the two coins in the lighter set, we can determine which is lighter. The lighter one is the counterfeit coin.

Example 1

Consider $a = 2$, $b = 2$, $T(1) = 2$, $f(n) = n$

$$
\begin{aligned}
T(n) &= aT(n/b) + f(n) \\
&= 2[T(n/2)] + n \\
&= 2[2T(n/4) + n/2)] + n \\
&= 4[T(n/4)] + 2n \\
&= 4[2T(n/8) + n/4] + 2n \\
&= 8T(n/8) + 3n
\end{aligned}
$$

$$
\vdots
$$

$$
T(n) = 2^i T(n/2^i) + in, \text{ for any } \log_2 n \geq i \geq 1.
$$
$$
T(n) = nT(1) + n \log_2 n = n \log_2 n + 2n
$$

Example 2

$a = 1$, $b = 2$, $f(n) = cn$, $c = $ constant

$$
\begin{aligned}
T(n) &= aT(n/b) + f(n) \\
&= 1T(n/2) + cn \\
&= T(n/2) + cn \\
&= \left[T(n/4) + \frac{cn}{2} \right] + cn \\
&= [T(n/4)] + \frac{3}{2}cn \\
&= \left[T(n/8) + \frac{cn}{4} \right] + \frac{3}{2}cn
\end{aligned}
$$

$$= T(n/8) + \frac{7}{4} cn$$

.
.

$$T(n) = T(n/2^i) + 2^i - 1 \frac{cn}{2^{i-1}}$$

EFFICIENCY ANALYSIS USING THE MASTER THEOREM

Theorem

The solution to the equation $T(N) = aT(N/b) + \theta(N^k)$, where $a \geq 1$ and $b > 1$ is

$$T(N) = \begin{cases} \theta\left(N^{\log_b^a}\right) & \text{if } a > b^k \\ \theta\left(N^k \log N\right) & \text{if } a = b^k. \\ \theta\left(N^k\right) & \text{if } a < b^k \end{cases}$$

This theorem can be used to determine the running time of most divide-and-conquer algorithms.

Example

Find the solution for the following recurrences:

1. $T(n) = 4T(n/2) + n$, $T(1) = 1$
   Compare with $T(n) = aT(n/b) + f(n)$.
   We get $a = 4$, $b = 2$, $f(n) = n$.
   $\theta(n^k) = n$
   $b^k = 2$
   $a > b^k$
   $$T(N) \in \theta\left(N^{\log_b^a}\right) \quad \text{if } a > b^k$$
   $$T(N) \in \theta\left(N^{\log_2^4}\right) \quad \text{if } a > b^k$$
   $$T(N) \in \theta\left(N^2\right)$$

2. $T(n) = 4T(n/2) + n^2$ $T(1) = 1$
   Compare with $T(n) = aT(n/b) + f(n)$.
   We get $a = 4$, $b = 2$, $f(n) = n^2$.
   $\theta(n^k) = n^2$
   $b^k = 2^2 = 4$
   $a = b^k$
   $$T(N) \in \theta\left(N^k \log N\right) \quad \text{if } a = b^k$$
   $$T(N) \in \theta(N^2 \log N)$$

3. $T(n) = 4T(n/2) + n^3 \ T(1) = 1$
   Compare with $T(n) = aT(n/b) + f(n)$.
   We get $a = 4, b = 2, f(n) = n^3$.
   $\theta(n^k) = n^3$
   $b^k = 2^3 = 8$
   a < b^k

   $$T(N) \in \theta(N^k) \quad \text{if } a < b^k$$

   $$T(N) \in \theta(N^3)$$

Divide-and-Conquer Examples
- Binary search
- Sorting: merge sort and quick sort
- Binary tree traversals
- Matrix multiplication: the Strassen algorithm
- Closest-pair algorithm
- Convex-hull algorithm
- Tower of Hanoi
- Exponentiation

2.2   BINARY SEARCH

A binary search is an efficient search method based on the divide-and-conquer algorithm. The binary search technique determines whether a given element is present in the list of sorted arrays. If we compare the target with the element that is in the middle of a sorted list, we have three possible results: the target matches, the target is less than the element, or the target is greater than the element. In the first and best case, we are done. In the other two cases, we learn that half of the list can be eliminated from consideration. When the target is less than the middle element, we know that if the target is in this ordered list, it must be in the list before the middle element. When the target is greater than the middle element, we know that if the target is in this ordered list, it must be in the list after the middle element. These facts allow this one comparison to eliminate one-half of the list from consideration. As the process continues, we will eliminate from consideration one-half of what is left of the list with each comparison.

Let an array a[i], $1 \le i \le n$ be a list of elements that are sorted in nondecreasing order (ascending order). The problem is to find whether the given element x is present in the list.

- If x is present then return the index 'j' such that a[j] = x.
- If x is not in the list, then j is to be set to zero.

Let $P = (n, a_i, a_{i+1}, \ldots, a_l, x)$ denote an arbitrary instance of this search problem.
Here n = number of elements in the list
$a_i, a_{i+1}, \ldots ; a_l$ = list of elements; and x = the element to be searched for.
Divide and conquer can be used to solve this problem.

Case 1

Let Small(P) be true if $n = 1$.
In this case, S(P) will take the value i if $x = a_i$.
Otherwise, it will take the value 0. Then, $g(n) = \theta(1)$.

Case 2

If P has more than one element, it can be divided (or reduced) into new subproblems as follows. Pick an index q in the range [i, l] and compare x with a[q].
There are three steps:

1. If $x = a[q]$, then the problem P is immediately solved.
2. If $x < a[q]$, then search x in the sub list a[i], a[I + 1], a[i + 2], ... a[q − 1].
   Now the problem P was reduced to $(q-i, a_i, \ldots, a_{q-1}, x)$
3. If $x > a[q]$, then search x in the sublist a[q + 1], a[q + 2], ..., a[l]. Now the problem P is reduced to $(1 - q, a_{q+1}, \ldots, a_1, x)$

Any given problem P gets divided (reduced) into one new subproblem. After a comparison with a[q], the instance remaining to be solved (if any) can be solved by using this divide-and-conquer scheme again. If q is always chosen such that a[q] is the middle element (i.e., $q = (n+1)/2$ or $q = (first + last)/2$), then the resulting search algorithm is known as binary search. Now the answer to the new subproblem is also the answer to the original problem 'P', so there is no need for any combining.

```
int BinSrch(Type a[], int i, int 1, Type x)
// Given an array a[i:l] of elements in nondecreasing order, l<=i<=l,
//determine whether x is present, and if so, return j such that x == a[j]; else
    return O.
{       if (l==i) {      //if Small(P)
        if (x==a[i]) return i;
        else return 0;
        }
        else { // Reduce P into a smaller subproblem.
        int mid = (i+l)/2;
        if (x == a[mid]) return mid;
        else if (x < a[mid]) return BinSrch(a,i,mid-1,x);
        else return BinSrch(a,mid+1,1,x);
        }
}
```

Program: Recursive Binary Search

```
int BinSearch(Type a[], int n, Type x)
// Given an array a[l:n] of elements in nondecreasing order, n>=0, determine
    whether
```

// x is present, and if so, return j such that x == a[j]; else return 0.

```
{
int low = 1, high = n;
while (low <= high){
int mid = (low + high)/2;
if (x < a[mid]) high = mid—1;
else if (x > a[mid]) low = mid + 1;
else return(mid);
}
return(0);
}
```

PROGRAM: ITERATIVE BINARY SEARCH

Example

Let us select the 14 entries −15, −6, 0, 7, 9, 23, 54, 82, 101, 112, 125, 131, 142, and 151; place them in a [1: 14] and simulate the steps that BinSearch goes through as it searches for different values of x. Only the variables low, high, and mid need to be traced when we simulate the algorithm.

We try the following values for x: 151, −14, and 9 for two successful searches and one unsuccessful search. The following table shows the traces of BinSearch on these three inputs.

x=151	low	high	mid		x=-14	low	high	mid
	1	14	7			1	14	7
	8	14	11			1	6	3
	12	14	13			1	2	1
	14	14	14			2	2	2
			found			2	1	not found

x=9	low	high	mid
	1	14	7
	1	6	3
	4	6	5
			found

Do likewise for all the numbers in the list and find the total number of comparisons.

Index	1	2	3	4	5	6	7	8	9	10	11	12	13	14
Element	−15	−6	0	7	9	23	54	82	101	112	125	131	142	151
Comparisons	3	4	2	4	3	4	1	4	3	4	2	4	3	4

Example: 2

Suppose k = 18 and we have the following array:

10, 12, 13, 14, 18, 20, 25, 27, 30, 35, 40, 45, 47

↑

Middle term

Solution

1. To obtain the middle element, we have to apply the formula:
   Mid = (low + high)/2
   Mid = (0 + 12)/2 = 6
   Mid = 6
2. Check a[6] = 25 with key. Then divide the array because k < 25 so we need to search the left array.
   10, 12, 13, 14, 18, 20

   Middle term
3. Conquer the sub-array by determining whether K is in the sub-array. This is accomplished by recursively dividing the sub-array. K is present in the left sub-array.
4. Compare K=18 with the middle element, K > 13. So K is present in the right sub-array.
   14, 18, 20
   ↑
   Middle term
5. Compare K with the middle element, K = 18. Yes. The element is present in the array.

ANALYSIS OF BINARY SEARCH ALGORITHM

Worst-Case Analysis

1. The search key is not in the sorted array.
2. The search key is in the last position.

Here the key comparison is done with half the size of the array.

Let T(n) be the worst-case number of comparisons used by procedure search on an array of n numbers. Suppose n is the power of 2:

$$T(n) = \begin{cases} 0 & if \; n = 1 \\ T(n/2) + 1 & otherwise \end{cases}$$

This recurrence is solved as

$$
\begin{aligned}
T(n) &= T(n/2) + 1 \\
&= (T(n/4) + 1) + 1 \\
&= T(n/4) + 2 \\
&= (T(n/8) + 1) + 2 \\
&= T(n/8) + 3 \\
&\;\;\vdots \\
&= T(n/2^i) + i \\
&\;\;\vdots \\
&= T(1) + \log n \quad (i.e., \; i = \log_2 n)
\end{aligned}
$$

$$= 0 + \log_2 n$$
$$T_w(n) = \log_2 n$$
$$T(n) = \theta(\log n)$$

Average-Case Analysis

The number of key comparisons in an average case is slightly smaller than that in the worst case. To obtain the average-case efficiency of a binary search, we will consider some samples of input n.

Input n	Total comparison C
1	1
2	2
4	3
8	4
16	5
.	.
.	.
.	.
128	8

Observing the given table, we can write

$$\log_2 n + 1 - C.$$
For instance if n = 2, then $\log_2 2 = 1$. Then, C = 1 + 1 = 2.
Similarly, n = 16; then, C = $\log_2 16 + 1$.
$$C = 4 + 1 = 5$$

Thus, we can write

$$T_{avg}(n) = \log_2 n + 1$$
$$T(n) = \theta(\log_2 n).$$

Best-Case Analysis

The number of comparisons used by procedure search on an array of n numbers is only one.

$$T_{best}(n) = 1$$
$$T(n) = \theta(1)$$

THEOREM

Function BinSearch (a, n, x) Works Correctly.

Proof

int BinSearch(Type a[], int n, Type x)
// Given an array a[l:n] of elements in nondecreasing order, n>=0, determine
   whether

// x is present, and if so, return j such that x == a[j]; else return 0.

```
{
int low = 1, high = n;
while (low <= high){
int mid = (low + high)/2;
if (x < a[mid]) high = mid—1;
else if (x > a[mid]) low = mid + 1;
else return(mid);
}
return(0);
}
```

We assume that all statements work as expected and that comparisons such as x > a [mid] are appropriately carried out.

Initially, low=1, high=n, n>=0, and a [1] <=a [2] <=. . . . .<=a[n].

If n==0, then the while loop is not entered, and 0 is returned. Otherwise, we observe that each time through the loop the possible elements to be checked for equality with x are a[low], a[low + 1] , . . . , a [mid] , . . . , a [high].
   If x==a [mid], then the algorithm terminates successfully. Otherwise, the range is narrowed by either increasing low to mid + 1 or decreasing high to mid − 1. Clearly this narrowing of the range does not affect the outcome of the search. If low becomes greater than high, then x is not present, and hence, the loop is exited.
   Suppose we begin by determining the time for BinSearch on the previous data set. We concentrate on comparisons between x and the elements in a [], recognizing that the frequency count of all other operations is of the same order as that for these comparisons. Comparisons between x and elements of a [] are referred to as *element comparisons*. The number of element comparisons needed to find each of the 14 elements is

Index	1	2	3	4	5	6	7	8	9	10	11	12	13	14
Element	−15	−6	0	7	9	23	54	82	101	112	125	131	142	151
Comparisons	3	4	2	4	3	4	1	4	3	4	2	4	3	4

No element requires more than four comparisons to be found.
   From the preceding table on successful search of an element 'x' in the array, it gives the following:

   Best case—1 comparison
   Worst case—4 comparisons
   Average case—3 comparisons

On unsuccessful search, there are 15 possible ways that an unsuccessful search may terminate depending on the value x.

- If x < a [1], then the algorithm needs three element comparisons to determine that x is not present.
- For all the remaining possibilities, BinSearch requires four element comparisons.
- Thus the average number of comparisons for an unsuccessful search is

$$= (3 + [14 * 4])/15 = 59/15 = 3.93.$$

Binary Decision Tree

As shown in Figure 2.2, the value of each node is the value of 'mid'. If n = 14, then the 'mid' values are produced by BinSearch. If 'x' is present, then the algorithm will end at one of the circular nodes that lists the index into the array where x was found. If x is not present, the algorithm will terminate at one of the square nodes. Circular nodes are called *internal* nodes, and square nodes are referred to as *external nodes*.

THEOREM

If n is in the range [2^{k-1}, 2k], then BinSearch makes at most k element comparisons for a successful search and either $k - 1$ or k comparisons for an unsuccessful search. (In other words, the time for a successful search is O(log n) and for an unsuccessful search is θ (log n)).

Proof

Consider the binary decision tree describing the action of BinSearch on n elements. All successful searches end at a circular node whereas all unsuccessful searches end at a square node.

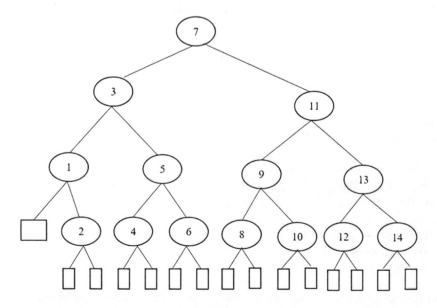

FIGURE 2.2   Binary decision tree for binary search, $n = 14$

If $2^{k-1} \le n < 2^k$, then all circular nodes are at levels 1, 2, ..., k whereas all square nodes are at levels k and $k + 1$ (note that the root is at level 1). The number of element comparisons needed to terminate at a circular node on level i is i whereas the number of element comparisons needed to terminate at a square node at level i is only i – 1.

This theorem states the worst-case time for binary search. To determine the average behavior, we need to look more closely at the binary decision tree and equate its size to the number of element comparisons in the algorithm.

The *distance* of a node from the root is one less than its level. The *internal path length* **I** is the sum of the distances of all internal nodes from the root. The *external path length* **E** is the sum of the distances of all external nodes from the root.

It is easy to show by induction that for any binary tree with n internal nodes, E and I are related by the formula

$$E = I + 2n.$$

It turns out that there is a simple relationship between **E**, **I**, and the average number of comparisons in binary search.

Let A_s **(n)** be the average number of comparisons in a successful search, and A_u **(n)** the average number of comparisons in an unsuccessful search. The number of comparisons needed to find an element represented by an internal node is one more than the distance of this node from the root. Hence,

$$A_s(n) - 1 + I/n.$$

The number of comparisons on any path from the root to an external node is equal to the distance between the root and the external node. Since every binary tree with n internal nodes has $n + 1$ external nodes, it follows that

$$A_u(n) = E/(n+1).$$

Using these three formulas for **E**, **A$_s$(n)**, and **A$_u$(n)**, we find that

$$A_u(n) = \frac{E}{(n+1)}$$

$$A_u(n) = \frac{1+2n}{(n+1)}$$

$$A_u(n) = \frac{n(A_s(n)-1)+2n}{(n+1)}$$

$$A_u(n) = \frac{n.A_s(n)-n+2n}{(n+1)}$$

$$A_u(n) = \frac{n.A_s(n)+n}{(n+1)}$$

$$A_u(n) = \frac{n(1 + A_s(n))}{(n+1)}$$

$$1 + A_s(n) = \frac{(n+1)A_u(n)}{n}$$

$$A_s(n) = \left(1 + \frac{1}{n}\right)A_u(n) - 1.$$

From this theorem, it follows that **E** is proportional to **n log n**. Using this in the preceding formulas, we conclude that $A_s(n)$ and $A_u(n)$ are both proportional to **log n**. Thus, we conclude that the average- and worst-case numbers of comparisons for a binary search are the same within a constant factor. The best-case analysis is easy. For a successful search only one element comparison is needed. For an unsuccessful search, the theorem states that $\log n$ element comparisons are needed in the best case.

In conclusion, we are now able to completely describe the computing time of binary search by giving formulas that describe the best, average, and worst cases:

Successful searches			Unsuccessful searches
$\theta(1)$	$\theta(\log n)$	$\theta(\log n)$	$\theta(\log n)$
Best	Average	Worst	Best, average, and worst

2.3   FINDING THE MAXIMUM AND MINIMUM

A finding-maximum-and-minimum problem is used to find the maximum item and minimum item in a given set of n elements.

A straightforward algorithm is given in the following program:

```
void StraightMaxMin(Type a[], int n, Type& max, Type& min)
// Set max to the maximum and min to the minimum of a[l:n].
{      max = min = a[l];
          for (int i=2; i<=n; i++) {
          if (a[i] > max) max a[i];
          if (a[i] < min) min = a[i];
          }

}
```

For example,

```
n = 5, a = [15, 20, 25, 30, 35]
max = min = a[1]=15
for i = 2 to 5
```

i = 2, if (a [2] > max), max = 20
i = 3, a[3] > 20 max = 25
i = 4, a[4] > 25 max = 30
i = 5, a[5] > 30 max = 35

Here we need four comparisons (n − 1).

The function StraightMaxMin requires **2(n − 1)** element comparisons in the best, average, and worst cases. An immediate improvement is possible by realizing that the comparison a [i] < min is necessary only when a [i] > max is false. Hence, we can replace the contents of the loop by

if (a[i] > max) max = a[i];
else if (a[i] < min) min = a[i];

Now the best case occurs when the elements are in increasing order. The number of element comparisons is $n − 1$. The worst case occurs when the elements are in decreasing order. In this case, the number of element comparisons is $2(n − 1)$.

A DIVIDE-AND-CONQUER ALGORITHM FOR THIS PROBLEM WOULD PROCEED AS FOLLOWS

Steps

1. Divide the array into sub-arrays of sizes 1 or 2.
2. Find their maximum and minimum.
3. Do it recursively to find the maximum and minimum of the given array.

Let P = (n, a[i], . . . , a[j]) denote an arbitrary instance of the problem. Here **n** is the number of elements in the list **a[i], . . . , a[j]**. We are interested in finding the maximum and minimum of this list.

- Let Small (P) be true when n ≤ 2. In this case, the maximum and minimum are **a[i]** if $n = 1$.
- If $n = 2$, the problem can be solved by making one comparison.
- If the list has more than two elements, **P** has to be divided into smaller instances.

For example, divide the problem **P** into two instances $P_1 = (\lfloor n/2 \rfloor, a [1], . . . ,$ a $[\lfloor n/2 \rfloor])$ and $P_2 = (n - \lfloor n/2 \rfloor, a [\lfloor n/2 \rfloor + 1] . . . a[n])$.

After subdividing the problem into smaller subproblems, we can solve them by recursively invoking the same divide-and-conquer algorithm, and finally, the results can be combined to give the main solutions.

If MAX (P) and MIN (P) are the maximum and minimum of the elements in P, then MAX (P) = larger value of MAX (P1) and MAX (P2).

Also, MIN (P) = smaller value of MIN (P1) and MIN (P2).

```
void MaxMin(int i, int j, Type& max, Type& min)
// a[l:n] is a global array. Parameters i and j are integers, 1 <= i <= j <= n.
   The effect //is to set max and min to the largest and smallest values in a[i:j],
   respectively.
{
      if (i == j) max = min = a[i]; // Small(P)
      else if (i == j–1) { // Another case of Small(P)
      if (a[i] < a[j]) {max a[j]; min = a[i]; }
      else {max = a[i]; min = a[j]; }
      }
      else {            // If P is not small
             // divide P into subproblems.
             // Find where to split the set.
      int mid=(i+j)/2; Type max1, min1;
             // Solve the subproblems.
      MaxMin(i, mid, max, min);
      MaxMin(mid+1, j, max1, min1);
             // Combine the solutions.
      if (max < max1) max = max1;
      if (min> min1) min = min1;
      }
}
```

MaxMin is a recursive function that finds the maximum and minimum of the set of elements {a(i), a(i + 1), . . . , a(j)}. The situation of set sizes one (i = j) and two (i = j − 1) are handled separately. For sets containing more than two elements, the midpoint is determined (just as in binary search) and two new subproblems are generated.

When the maxima and minima of these subproblems are determined, the two maxima are compared and the two minima are compared to achieve the solution for the entire set.

The procedure is initially invoked by the statement

MaxMin (1, n, x, y)

Suppose we simulate the function MaxMin on the following nine elements:

[1]	[2]	[3]	[4]	[5]	[6]	[7]	[8]	[9]
22	13	−5	−8	15	60	17	31	47

A good way of keeping track of recursive calls is to build a tree by adding a node each time a new call is made. For this program, each node has four items of information: i, j, max, and min.

We see that the root node contains 1 and 9 as the values of i and j corresponding to the initial call to MaxMin. This execution produces two new calls to MaxMin, where i and j have the values 1, 5 and 6, 9, respectively, and, thus, split the set into two

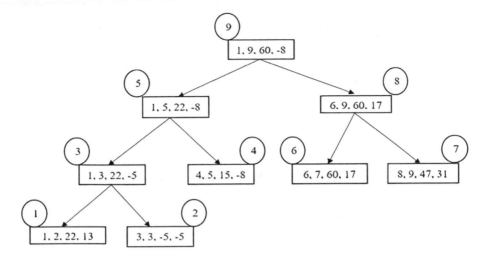

FIGURE 2.3   Trees of recursive calls of MaxMin

subsets of approximately the same size. From the tree, we can immediately see that the maximum depth of recursion is four (including the first call). The circled numbers shown in Figure 2.3 in the upper left corner of each node represent the orders in which max and min are assigned values.

WHAT IS THE NUMBER OF ELEMENT COMPARISONS NEEDED FOR MAXMIN?

If **T(n)** represents this number, then the resulting recurrence relation is

$$T(n) = \begin{cases} T\left(\dfrac{n}{2}\right) + T\left(\dfrac{n}{2}\right) + 2 & n > 2 \\ 1 & n = 0, \\ 0 & n = 1 \end{cases}$$

where n = powers of 2,

$$= 2^k \; k = \text{positive integer.}$$
$$T(n) = 2T(n/2) + 2$$
$$= 2[2T(n/4) + 2] + 2$$
$$= 4T(n/4) + 4 + 2$$
$$= 8T(n/8) + 8 + 4 + 2$$
$$\vdots$$

$$T(n) = 2^{k-1} T\left(\frac{n}{2^{k-1}}\right) + \sum_{1 \le i \le k-1} 2^i$$

$$= 2^{k-1} T\left(\frac{n}{2^{k-1}}\right) + \sum_{1 \le i \le k-1} 2^i$$

Put n = 2^k.

$$= 2^{k-1} T\left(\frac{2^k}{2^{k-1}}\right) + \sum_{1 \le i \le k-1} 2^i$$

$$= 2^{k-1} T(2) + \sum_{1 \le i \le k-1} 2^i$$

$$= 2^{k-1} T(2) + \sum_{1 \le i \le k-1} 2^i \; where\, T(2) = 1$$

$$= 2^{k-1} + \frac{2\left(2^{(k-1)} - 1\right)}{2-1} = 2^{(k-1)+1} - 2$$

$$= 2^{k-1} + 2^k - 2$$

$$= \frac{2^k}{2} + 2^k - 2$$

$$= \left(\frac{n}{2}\right) + n - 2 \quad where\, n = 2^k$$

$$T(n) = \frac{3n}{2} - 2$$

$\frac{3n}{2} - 2$ is the best-, average-, and worst-case number of comparisons when n is a power of 2; that is, n = 2^k.

2.4 MERGE SORT

Merge sort is one of the earliest algorithms proposed for sorting. According to Knuth, it was suggested by John von Neumann as early as 1945. Merging is the process of combining two or more files into a new sorted file. Merge sort is based on the idea that merging two sorted lists can be done quickly. Because a list with just one element is sorted, merge sort will break a list down into one-element pieces and then sort as it merges those pieces back together. All the work for this algorithm, therefore, occurs in the merging of the two lists.

Merge sort can be written as a recursive algorithm that does its work on the way up in the recursive process. In looking at the algorithm that follows, you will notice that it breaks the list in half as long as the first is less than the last. When we get to a point where the first and the last are equal, we have a list of one element, which is inherently sorted. When we return from the two calls to Merge Sort that have lists of size 1, we then call Merge Lists to put those together to create a sorted list of size 2. At the next level up, we will have two lists of size 2 that get merged into one sorted list of size 4. This process continues until we get to the top call, which merges the two sorted halves of the list back into one sorted list. We see that merge sort breaks a list in halves on the way down in the recursive process and then puts the sorted halves together on the way back up.

- It is a perfect example of the successful application of the divide-and-conquer technique.
- We assume that the elements are to be sorted in nondecreasing order.
- Given a sequence of n elements (also called keys) a[1], ..., a[n], the general idea is to imagine them split into two sets a[1], ..., a[n/2] and a[n/2+1], ..., a[n].
- Each set is individually sorted, and the resulting sorted sequences are merged to produce a single sorted sequence of n elements.

DIVIDE-AND-CONQUER STEPS OF MERGE SORT

1. Divide the array into two sub-arrays each with n/2 items.
2. Conquer each sub-array by sorting it unless the array is sufficiently small; use recursion to do this.
3. Combine the solutions to the sub-arrays by merging them into a single-sorted array.

Example
Consider the array of ten elements a [1: 10] = (310, 285, 179, 652, 351, 423, 861, 254, 450, 520).

- Function Merge Sort begins by splitting a [ ] into two sub-arrays each of size 5 (a [1: 5] and a [6: 10]).
- The elements in a [1: 5] are then split into two sub-arrays of size 3 (a [1: 3]) and 2 (a [4: 5]).
- Then the items in a [1: 3] are split into sub-arrays of size 2 (a [1: 2]) and 1 (a [3: 3]).
- The two values in a [1: 2] are split a final time into one-element sub-arrays, and now the merging begins.
- Note that no movement of data has yet taken place.
- A record of the sub-arrays is implicitly maintained by the recursive mechanism.

Pictorially the file can now be viewed as

(310 | 285 | 179 | 652, 351 | 423, 861, 254, 450, 520).

Vertical bars indicate the boundaries of the sub-arrays.

- Elements a[1] and a[2] are merged to yield

(285, 310 | 179 | 652, 351 | 423, 861, 254, 450, 520).

- Then a[3] is merged with a[1:2]:

   (179, 285, 310| 652, 351 | 423, 861, 254, 450, 520).

- Next, elements a[4] and a[5] are merged:

   (179, 285, 310| 351, 652 | 423, 861, 254, 450, 520).

- Next a[1:3] and a[4:5] are merged:

   (179, 285, 310, 351, 652| 423, 861, 254, 450, 520).

At this point, the algorithm has returned to the first invocation of Merge Sort and is about to process the second recursive call. Repeated recursive calls are invoked producing the following sub-arrays:

   (179, 285, 310, 351, 652| 423 | 861 | 254 | 450, 520)

- Elements a[6] and a[7] are merged.

   (179, 285, 310, 351, 652 | 423, 861 | 254 | 450, 520)

- Then a[8] is merged with a[6:7].

   (179, 285, 310, 351, 652 | 254, 423, 861 | 450, 520)

- Next, a[9] and a[10] are merged.

   (179, 285, 310, 351, 652 | 254, 423, 861 | 450, 520)

- Next a[6:8] and a[9:10] are merged.

   (179, 285, 310, 351, 652 | 254, 423, 450, 520, 861)

At this point, there are two sorted sub-arrays, and the final merge produces the fully sorted result:

   (179, 254, 285, 310, 351, 423, 450, 520, 652, 861).

Figure 2.4 is a tree that represents the sequence of recursive calls that are produced by Merge Sort when it is applied to ten elements. The pair of values in each node are the values of the parameters low and high. Notice how the splitting continues until sets containing a single element are produced.

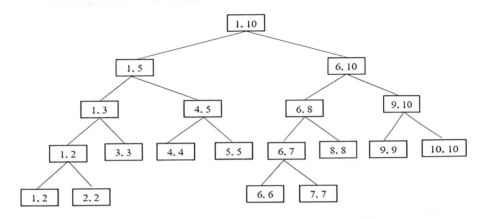

FIGURE 2.4   Tree of calls of Merge Sort (1, 10)

For example, the node containing 1, 2, and 3 represents the merging of a [1:2] with a [3]. The program for merge sort is given as follows:

```
void MergeSort(int low, int high)
// a[low : highJ is a global array to be sorted.Small(P) is true if there is only
   one element //to sort. In this case the list is already sorted.
{
         if (low < high) [ // If there are more than one element
                  // Divide P into subproblems.
                  // Find where to split the set.
         int mid = (low + high)/2;
                  //Solve the subproblems.
         MergeSort(low, mid);
         MergeSort(mid + 1, high);
                  // Combine the solutions.
         Merge(low, mid, high);
         }
}
```

The program for merging two sorted sub-arrays using auxiliary storage is given as

```
void Merge(int low, int mid, int high)
// a[low:high] is a global array containing two sorted subsets in a[low:mid]
//and in a[mid+1:high]. The goal is to merge these two sets into a single set
//residing in a[low:high]. b[] is an auxiliary global array.
{
         int h = low, i = low, j = mid+1, k;
         while ((h <= mid) && (j <= high)) {
```

```
if (a[h] <= a[j]) {b[i]= a[h]; h++; }
else { b[i] = a[j]; j++; } i++;
}
if (h > mid) for (k=j; k<=high; k++) {
b[i] = a[k]; i++;
}
else for (k=h; k<=mid; k++) {
b[i] = a[k]; i++;
}
for (k=low; k<=high; k++) a[k] b [k] ;
}
```

If the time for the merging operation is proportional to n, then the computing time for the merge sort is described by the recurrence relation:

$$T(n) = \begin{cases} 0 & n = 1, \ a \ a \ constant \\ 2T\left(\dfrac{n}{2}\right) + cn & n > 1, \ c \ a \ contant. \end{cases}$$

When n is a power of 2, $\mathbf{n = 2^k}$, we can solve this equation by successive substitutions:

$$
\begin{aligned}
T(n) &= 2(2T \ (n/4) + cn/2) + cn \\
&= 4 \ T \ (n/4) + 2cn \\
&= 4(2T \ (n/8) + cn/4) + 2cn \\
&\quad \vdots \\
&= 2^k T\left(\frac{n}{2^k}\right) + kcn \\
&= 2^k T\left(\frac{2^k}{2^k}\right) + kcn \\
&= 2^k T(1) + kcn \qquad [T(1) = a][n = 2^k, \text{ so } k = \log_2 n] \\
T(n) &= an + cn \log_2 n \Rightarrow T(n) = O(n \log n)
\end{aligned}
$$

2.5 GREEDY ALGORITHMS: GENERAL METHOD

- Greedy algorithms are simple and straightforward. They take decisions on the basis of the information at hand without worrying about the effect these decisions may have in the future.
- They are easy to invent, easy to implement, and, most of the time, quite efficient.
- Greedy algorithms are used to solve optimization problems.
- Greedy method is the most important design technique, which makes a choice that looks best at that moment.
- Given 'n' inputs and we are required to form *subset*, such that it satisfies some given constraints, then such a subset is called *feasible solution*.

- A feasible solution that either maximizes or minimizes a given *objective function* is called an *optimal solution*.
- The greedy method suggests an algorithm that works in different stages but considers one input at a time.
- At each stage, a decision is made regarding whether a particular input is in an *optimal solution*. This is done by considering the inputs in an order determined by some *selection procedure*.
- The selection procedure itself is based on some optimization measure. This measure may be the *objective function*.
- If the inclusion of the next input into the partially constructed optimal solution will result in an infeasible solution, then this input is not added to the partial solution. Otherwise, it is added.
- There are several different optimization measures are present. Most of these will result in algorithms that generate suboptimal solutions. This version of the greedy technique is called the *subset paradigm*.

```
SolType Greedy(Type a[], int n)
// a[l:n] contains the n inputs.
{
        SolType solution = EMPTY; // Initialize the solution.
        for (int i=l; i<=n; i++) {
        Type x - Select(a);
        if Feasible(solution, x)
        solution = Union(solution, x);
        }
return solution;
}
```

- The function **Select** selects an input from a [] and removes it. The selected input's value is assigned to x.
- **Feasible** is a Boolean-valued function that determines whether x can be included in the solution vector.
- The function **Union** combines x with the solution and updates the objective function.
- The function **Greedy** describes the essential way that a greedy algorithm will look, once a particular problem is chosen and the functions **Select**, **Feasible**, and **Union** are properly implemented.

CHARACTERISTICS OF A GREEDY ALGORITHM

1. To solve a problem in an optimal way construct the solution from the given set of objects.
2. As the algorithm proceeds, two other sets get accumulated. Among this, one set contains the candidates that have been already considered and chosen while the other set contains the candidates that have been considered but rejected.

3. There is a function that checks whether the particular set of candidates provides a solution to our problem by ignoring the questions of optimality for
   the time needed.
4. The second function checks whether the set of candidates is feasible or not.
5. The third function is a selection function that indicates at any time, which
   of the remaining candidates is the most promising one.
6. An objective function that does not appear explicitly gives the value of a
   solution.

STRUCTURE OF GREEDY ALGORITHM

- Initially the set of chosen items is empty.
- At each step
  - item will be added in a solution set by using the selection function.
  - If the set would no longer be feasible, then
  - reject items under consideration.
  - Else, if set is still feasible, then
  - add the current item.

APPLICATIONS OF GREEDY STRATEGY

- Change making for normal coin denominations
- Minimum spanning tree
- Single-source shortest paths
- Simple scheduling problems
- Huffman codes
- Traveling salesperson problem
- Knapsack problem
- Cassette filling
- Bin packing
- Container loading
- Optimal merge

ADVANTAGE

- Greedy algorithms work fast when they work simple algorithms.
- They are easy to implement.

DISADVANTAGE

- Greedy algorithms don't always work.
- It is hard to prove correct.

Example 1: Change Making

A set of coins will be given. We are asked to select least number of coins from a
given amount.

for example, 100, 25, 10, 5, 1 (unlimited supply)

m = 289

Optimal solution = 2 × 100 + 3 × 25 + 1 × 10 + 4 × 1

ALGORITHM

```
function makechange(n)
{
        C = {100, 25, 10, 5, 1}
        S = 0 (Solution set)
        S₁ = 0 (Sum of item in S)
        while S₁≠n do
        x = largest item in C such that Sᵢ + x ≤ n.
        if there is no such item return failure.
        S← S₀ (a coin of value x)
        S₁ = S₁+x;
        return S;
}
```

- For example, a child buys candy valued at less than \$1 and gives a \$1 bill to the cashier.
- The cashier wishes to return the change using the fewest number of coins. Assume that an unlimited supply of quarters, dimes, nickels, and pennies is available.
- The cashier constructs the change in stages. At each stage, a coin is added to the change.
- This coin is selected using the greedy criterion: *At each stage, increase the total amount of change constructed by as much as possible.*
- To assure feasibility (i.e., the change given exactly equals the desired amount) of the solution, the selected coin should not cause the total amount of change given so far to exceed the final desired amount.
- Suppose that 67 cents in change is due to the child. The first two coins selected are quarters.
- A quarter cannot be selected for the third coin because such a selection results in an infeasible selection of coins (change exceeds 67 cents).
- The third coin selected is a dime, then a nickel is selected, and finally two pennies are added to the change.

Example 2: Machine Scheduling

You are given n tasks and an infinite supply of machines on which these tasks can be performed. Each task has a start time S_i and a finish time f_i; $S_i < f_i$, $[S_i, f_i]$ is the processing interval for task i. Two tasks I and j overlap iff their processing intervals overlap at a point other than the interval start or end.

For example, the interval [1, 4] overlaps with [2, 4] but not with [4, 7]. A feasible task-to-machine assignment is an assignment in which the tasks should not overlap. Therefore, in a feasible assignment, each machine works on at most one task

task	a	b	c	d	e	f	g
start	0	3	4	9	7	1	6
finish	2	7	7	11	10	5	8

FIGURE 2.5   Seven tasks

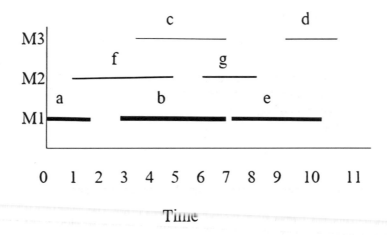

FIGURE 2.6   Schedule

at any time. An optimal assignment is a feasible assignment that utilizes the fewest number of machines.

Suppose we have $n = 7$ tasks labeled **a** through **g** and that their start and finish times are as shown in Figure 2.5.

A greedy way to obtain an optimal task assignment is to assign the tasks in stages. From the given set of jobs and machines, allocate jobs to a machine in which one is free as shown in Figure 2.6. Once a machine has been allocated, then it is called old machine. If a machine is not old, it is new.

For machine selection, use the greedy criterion: **If an old machine becomes available by the start time of the task to be assigned, assign the task to this machine; if not, assign it to a new machine.**

2.6   CONTAINER LOADING

A large ship is to be loaded with cargo. The cargo is containerized, and all containers are the same size. Different containers may have different weights. Let W_i be the weight of the ith container, $1 \le i \le n$, The cargo capacity of the ship is c. The problem is to load the ship with the maximum number of containers.

This problem can be formulated as an optimization problem in the following way:

Let x_i be a variable whose value can be either 0 or 1. If we set x_i to 0, then container **i** is not to be loaded. If x_i is 1, then the container is to be loaded.

Assign values to the x_i's that satisfy the constraints

$$\sum_{i=1}^{n} W_i X_i \leq c \text{ and } X_i \in \{0,1\}, 1 \leq i \leq n.$$

The optimization function is $\sum_{i=1}^{n} X_i$. Every set of x_i's that satisfies the constraints is a feasible solution. Every feasible solution that maximizes $\sum_{i=1}^{n} X_i$ is an optimal solution.

The ship may be loaded in stages, one container per stage. At each stage, decide which container to load.

GREEDY SOLUTION

- *From the remaining containers, select the one with the least weight.* This order of selection will keep the total weight of the selected containers minimum and, hence, leave the maximum capacity for loading more containers.
- Select the container that has the least weight, then the one with the next-smallest weight, and so on until either all containers have been loaded or there isn't enough capacity for the next one.

GREEDY CONTAINER LOADING ALGORITHM

1. Sort the containers in increasing order of their weights.
2. Load the container that has the least weight.
3. Find the total weight of the container loader.
4. If the total weight of the container is less than the capacity **c**, then repeat steps 2 to 4.

PROGRAM FOR LOADING CONTAINERS IS GIVEN AS

```
void ContainerLoading(Container* c, int capacity, int numberOfContainers,
    int* x)
{
// Greedy algorithm for container loading. Set x[i] = 1 iff container i, i > = 1
    is loaded.
// sort into increasing order of weight
        Sort(c, numberOfContainers);
        int n = numberOfContainers;
        // initialize x
        for (int i = 1; i <= n; i++)
        x [i] = 0;
        //select containers in order of weight
        for (int i = 1; i <= n && c[i] .weight <= capacity; i++)
        {
        // enough capacity for container c[i] .id
```

```
        x [c[i] .id] = 1;
        capacity -= c[i] .weight; // remaining capacity
        }
    }
```

Example

Suppose that $n = 8$, $[w_1, w_2, w_3, \ldots w_8] = [100, 200, 50, 90, 150, 50, 20, 80]$ and $c = 400$.

Solution

When the greedy algorithm is used, the containers are considered for loading in the order 7, 3, 6, 8, 4, 1, 5, 2.

$$[w_7, w_3, w_6, w_8, w_4, w_1, w_5, w_2] = [20, 50, 50, 80, 90, 100, 150, 200]$$

Containers 7, 3, 6, 8, 4, and 1 together weigh 390 units and are loaded. The available capacity is now 10 units, which is inadequate for any of the remaining containers.

In the greedy solution, we have $[x_1, x_2, x_3, \ldots, x_8] = [1, 0, 1, 1, 0, 1, 1, 1]$ and $\sum X_i = 6$.

ANALYSIS OF THE ALGORITHM

- Sorting the container takes $O(n \log n)$.
- The remainder of the algorithm takes $O(n)$ time.
- The overall time complexity of the algorithm is $O(n \log n)$.

2.7   KNAPSACK PROBLEM

The greedy method is best suited to solving more complex problems such as a knapsack problem. A bag or knapsack is given with capacity **m** and **n** objects. Object **i** has a weight w_i and profit P_i. A fraction of the object is considered as x_i, where $0 \leq x_i \leq 1$. If this fraction $x = 1$, then the entire object is put into sack. When we place this fraction into the sack, we get $w_i x_i$ and $P_i x_i$.

OBJECTIVE FUNCTION

The objective is to obtain a filling of the knapsack that maximizes the total profit earned. Since the knapsack capacity is m, we require the total weight of all chosen objects to be at most m. formally, the problem can be stated as

$$\sum_{1 \leq i \leq n} P_i x_i \tag{1}$$

$$\text{subject to } \sum_{1 \leq i \leq n} w_i x_i \leq m \tag{2}$$

$$\text{and } 0 \leq x_i \leq 1, 1 \leq i \leq n. \tag{3}$$

Any solution which satisfies **1** and **3** is called a feasible solution. Any feasible solution that satisfies **2** is called an optimal solution.

The greedy knapsack problem was divided into three categories:

1. Consider the objects with the largest profit values.
2. Consider the objects in nondecreasing weights.
3. Find $\dfrac{P_i}{W_i}$ and select objects starts from highest $\dfrac{P_i}{W_i}$ to the lowest.

Example: Consider the following instance of the knapsack problem:

$$n = 3,\ m = 20,\ (P_1, P_2, P_3) = (25, 24, 15)\ \text{and}\ (W_1, W_2, W_3) = (18, 15, 10)$$

SOLUTION

1. Choose the object that has the highest profit.

$$P_1 = 25 \qquad W_1 = 18$$

Now we take W_1 completely since we have a capacity of 20. $X_1 = 1$

$$X_1 W_1 = 1 \times 18 = 18 \qquad X_1 P_1 = 1 \times 25 = 25$$

The remaining weight = 2, so try the second object.

$$X_2 = \frac{2}{15}\, i.e. \left[\frac{20 - 18}{P_2} \right]$$

$$X_2 W_2 = \frac{2}{15} \times 15 = 2 \qquad X_2 P_2 = \frac{2}{15} \times 24 = 3.2$$

X_1	X_2	X_3	$\sum x_i w_i$	$\sum x_i P_i$
1	2/15	0	20	28.2

2. Choose some other object that has the minimum weight.

$$X_1 = 1, \qquad X_1 W_1 = 1 \times 10 = 10 \qquad X_1 P_1 = 1 \times 15 = 15$$

$$X_2 = \frac{10}{15} = \frac{2}{3} \qquad X_2 W_2 = \frac{2}{3} \times 15 = 10 \qquad X_2 P_2 = \frac{2}{3} \times 24 = 16$$

X_1	X_2	X_3	$\sum x_i w_i$	$\sum x_i P_i$
1	2/3	0	20	31

3. To get the highest profit, do the following method:

1. $\dfrac{P_1}{W_1} = \dfrac{25}{18} = 1.38$

2. $\dfrac{P_2}{W_2} = \dfrac{24}{15} = 1.6$

3. $\dfrac{P_3}{W_3} = \dfrac{15}{10} = 1.5$

So select the value of the highest ratio; now the second one is the highest value.

$$X_1 = 1, X_1 W_2 = 1 \times 15 = 15 \qquad X_1 P_2 = 1 \times 24 = 24$$

Then $X_2 = \dfrac{5}{10} = \dfrac{1}{2}$

$$X_2 W_3 = \dfrac{1}{2} \times 10 = 5 \qquad X_2 P_3 = \dfrac{1}{2} \times 15 = 7.5.$$

X_1	X_2	X_3	$\sum x_i w_i$	$\sum x_i P_i$
1	½	0	20	31.5

This is the optimal solution.

Method 1

- Try to fill the sack by considering an object with the largest profit.
- If an object under consideration does not fit into the sack, then a fraction of it is included.
- Thus, each time an object is included in the knapsack, we obtain the largest possible increase in profit value.
- Thus, each time an object is included in the knapsack, the profit increases.

Method 2

- Consider the objects in the order of nondecreasing weights (i.e., choose the object that has the lowest weight) to fill the knapsack first. Since this fits into the sack completely, $X_1 = 1$.
- Choose the next-lowest weight to fill into the knapsack. Since this size cannot be completely included in sack, take a fraction of it.
- This method is also suboptimal only.

Method 3

- This algorithm finds the values between a rate at which the profit increases and the rate at which the capacity is used.

- At each step, we include the object that has the maximum profit per unit of capacity used.
- It means that the objects are considered in the order of ratio (P_i/W_i). So the object with the highest ratio is added first.
- This method gives the optimal solution.

void GreedyKnapsack(float m, int n)
//p[1:n] and w[1:n] contain the profits and weights, respectively, of the n objects
// ordered such that p[i]/w[i] >= p[i+1]/w[i+1]. m is the knapsack size and
//x[1:n] is the solution vector.

```
{
        for (int i=l; i<=n; i++) x[i] = 0.0; // Initialize x.
        float U = m;
        for (i=l; i<=n; i++) {
        if (w[i] > U) break;
        x [i] = 1. 0;
        U -= w [i] ;
        }
        if (i <= n) x [ i]= U/w[i] ;
}
```

There are two versions of this problem:

1. Fractional knapsack
2. 0–1 knapsack

FRACTIONAL KNAPSACK PROBLEM

The setup is the same, but the items can be broken into smaller pieces so that only a fraction of x_i of item **i**, where $0 \le x_i \le 1$.

0–1 Knapsack Problem

The setup is the same, but the items may not be broken into smaller pieces, so an item may be taken or left but a fraction of an item cannot be taken.

The fractional knapsack problem can be solved by the greedy strategy whereas the 0–1 problem cannot.

3 Dynamic Programming

3.1 INTRODUCTION: DYNAMIC PROGRAMMING

Dynamic programming is an algorithm design technique with a rather interesting history. It was invented by a prominent U.S. mathematician, Richard Bellman, in the 1950s as a general method for optimizing multistage decision processes. Thus, the word *programming* in the name of this technique stands for 'planning' and does not refer to computer programming. After proving its worth as an important tool of applied mathematics, dynamic programming has eventually come to be considered, at least in computer science circles, as a general algorithm design technique that does not have to be limited to special types of optimization problems. It is from this point of view that we consider this technique here.

Dynamic programming is a method in which the solution to a problem is obtained from the sequence of decisions. The optimal solution to the given problem is obtained from the sequence of all possible solutions being generated. Dynamic programming works in the same way as divide and conquer by combining solutions to subproblems.

- Divide and conquer partitions a problem into independent subproblems.
- The greedy method only works with local information.

CONDITION TO SOLVE BY THE GREEDY METHOD

Condition: an optimal sequence of decisions can be found by making the decisions one at a time and never making an erroneous decision.

CONDITION TO SOLVE BY THE DYNAMIC PROGRAMMING

Condition: an optimal sequence of decisions can be found by making many decisions at a time and all the overlapping sub-instances are considered.

DYNAMIC PROGRAMMING: GENERAL METHOD

Dynamic programming is an algorithm design method that can be used when the solution to a problem can be viewed as the result of a sequence of decisions. The problems that can be solved in which optimal decisions can be made once at a time are

- the knapsack problem,
- optimal merge patterns, and
- the shortest path.

DOI: 10.1201/9781003355403-3

Example 3.1 Knapsack

The solution to the knapsack problem can be viewed as the result of a sequence of decisions. We have to decide the values of Xi, $1 \leq I \leq n$. First, we make a decision on X_1, then on X_2, then on X_3, and so on. An optimal sequence of decisions maximizes the objective function $\sum PiXi$.

Example 3.2 Optimal Merge Patterns

An optimal merge pattern tells us which pair of files should be merged at each step. The *solution* to an optimal merge pattern using dynamic programming is to decide which pair of files should be merged first, which pair second, which pair third, and so on. An optimal sequence of decisions is a least-cost sequence.

Example 3.3 Shortest Path

One way to find the shortest path from vertex i to vertex j in a directed graph G is to decide which vertex should be the second vertex, which the third, which the fourth, and so on until vertex j is reached. An optimal sequence of decisions is one that results in a path of shortest length.

Example 3.4 Shortest Path

Suppose we wish to find the shortest path from vertex i to vertex j. Let Ai be the vertices adjacent from vertex i. Which of the vertices in Ai should be the second vertex on the path? There is no way to make a decision at this time and guarantee that future decisions leading to an optimal sequence can be made. If, however, we wish to find the shortest path from vertex i to all other vertices in G, then at each step, a correct decision can be made.

One way to solve problems for which it is not possible to make a sequence of stepwise decisions that lead to an optimal decision sequence is to try out all possible decision sequences. We could enumerate all decision sequences and then pick out the best. But the time and space requirements may be prohibitive.

Dynamic programming often drastically reduces the amount of enumeration by avoiding the enumeration of some decision sequences that cannot possibly be optimal. In dynamic programming, an optimal sequence of decisions is arrived at by making an explicit appeal to the *principle of optimality*.

Definition (Principle of Optimality)

The principle of optimality states that an optimal sequence of decisions has the property that whatever the initial state and decision are, the remaining decisions must constitute an optimal decision sequence with regard to the state resulting from the first decision.

Example: Shortest path and 0/1 knapsack.

Greedy Method Versus Dynamic Programming

Greedy method: In the greedy method, only one decision sequence is ever generated.

Dynamic programming:

- In dynamic programming, many decision sequences may be generated.
- However, sequences containing suboptimal subsequences are discarded because they cannot be optimal due to the principle of optimality.

Features of Dynamic Programming

1. It is easier to obtain the recurrence relation using a backward approach.
2. Dynamic programming algorithms often have polynomial complexity.
3. Optimal solutions to subproblems are retained so as to avoid recomputing their values.

3.2 MULTISTAGE GRAPHS

A multistage graph $G = (V, E)$ is a directed graph in which the vertices are partitioned into $k \geq 2$ disjoint sets Vi, $1 \leq i \leq k$. In addition, if <u, v> is an edge in E, then $u \in V_i$ and $v \in V_{i+1}$ for some i, $1 \leq i \leq k$. The sets V_1 and V_k are such that $|V_1| = |V_k| = 1$.

Let s and t, respectively, be the vertex in V_1 and V_k. The vertex s is the *source*, and t the *sink*. Let c(i, j) be the cost of edge (i, j). The multistage graph problem is to find a minimum cost path from **s** to **t**. The cost of a path from **s** to **t** is the sum of the costs of the edges on the path from **s** to **t**. Figure 3.1 shows a five-stage graph.

A dynamic programming formulation for a **k-stage** graph problem is obtained by two approaches:

1. Forward
2. Backward

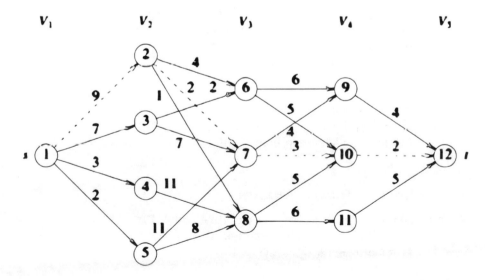

FIGURE 3.1   Five-stage graph

Forward Approach

A dynamic programming formulation for a **k-stage** graph problem is obtained by first noticing that every *s*-to-*t* path is the result of a sequence of $k - 2$ decisions. The **i**th decision involves determining which vertex in V_{i+1} $1 \leq i \leq k - 2$, is to be on the path. It is easy to see that the principle of optimality holds.

Let **p(i, j)** be a minimum-cost path from vertex **j** in V_i to vertex **t**. Let *cost(i, j)* be the cost of this path. Then using the forward approach, we obtain

$$cost(i, j) = \min\{c(j, l) + cost(i+1, l)\}$$

$$l \in V_{i+1}$$

$$\langle j, l \rangle \in E.$$

Since $cost(k-1, j) = c(j, t) if \langle j, t \rangle \in E$ and $cost(k-1, j) = \infty$ if $\langle j, t \rangle \notin E$.

Multistage-Graph Pseudo Code Corresponding to the Forward Approach

```
void FGraph(graph G, int k, int n, int p[])
// The input is a k-stage graph G = (V,E) with n vertices indexed in order of
    stages.
// E is a set of edges and c[i] [j] is the cost of <i,j>. p[1:k] is a minimum-cost
    path.
{
        float cost [MAXSIZE]; int d[MAXSIZE], r;
        cost[n] = 0.0;
        for (int j=n-1; j>=1; j--) {          // Compute cost[j].
        let r be a vertex such that <j,r> is an edge
        of G and c[j] [r] + cost[r] is minimum;
        cost [j] = c [j] [r] + cost [r] ;
        d[j] = r;
        }
        // Find a minimum-cost path.
        p[1] = 1; p[k] = n;
        for (j=2; j <= k-1; j++) p[j]= d[p[j-1]] ;
}
```

Solution Using the Forward Approach

$$
\begin{aligned}
Cost(3, 6) &= \min\{c(6, 9) + cost(4, 9), c(6, 10) + cost(4, 10)\} \\
&= \min\{6 + cost(4, 9), 5 + cost(4, 10)\} \\
&= \min\{6 + c(9, 12), 5 + c(10, 12)\} \\
&= \min\{6 + 4, 5 + 2\} \\
&= \min\{10, 7\} \\
&= 7
\end{aligned}
$$

Cost(4, 9) = 4 from $cost(k-1, j) = c(j,t) \, if \, \langle j,t \rangle \in E$

Cost(4, 9) = c(9, 12) = 4

Cost(3, 7) = min{c(7, 9) + cost(4, 9), c(7, 10) + cost(4, 10)}
$\qquad$ = min{4 + cost(4, 9), 3 + cost(4, 10)}
$\qquad$ = min{4 + c(9, 12), 3 + c(10, 12)}
$\qquad$ = min{4 + 4, 3 + 2}
$\qquad$ = min{8, 5}
$\qquad$ = 5

Cost(3, 8) = min{c(8, 10) + cost(4, 10), c(8, 11) + cost(4, 11)}
$\qquad$ = min{5 + cost(4, 10), 6 + cost(4, 11)}
$\qquad$ = min{5 + c(10, 12), 6 + c(11, 12)}
$\qquad$ = min{5 + 2, 6 + 5}
$\qquad$ = min{7, 11}
$\qquad$ = 7

Cost(2, 2) = min{c(2, 6) + cost(3, 6), c(2, 7) + cost(3, 7), c(2, 8) + cost(3, 8)}
$\qquad$ = min{4 + 7, 2 + 5, 1 + 7}
$\qquad$ = min{11, 7, 8} = 7

Cost(2, 3) = min{c(3, 6) + cost(3, 6), c(3, 7) + cost(3, 7)}
$\qquad$ = min{2 +cost(3, 6), 7 + cost(3, 7)}
$\qquad$ = min{2 + 7, 7 + 5}
$\qquad$ = min{9, 12}
$\qquad$ = 9

Cost(2, 4) = min{c(4, 8) + cost(3, 8)}
$\qquad$ = min{11 + 7}
$\qquad$ = 18

Cost(2, 5) = min{c(5, 7) + cost(3, 7), c(5, 8) + cost(3, 8)}
$\qquad$ = min{11 + cost(3, 7), 8 + cost(3, 8)}
$\qquad$ = min{11 + 5, 8 + 7}
$\qquad$ = min {16, 15}
$\qquad$ = 15

Cost(1, 1) = min{c(1, 2) + cost(2, 2), c(1, 3) + cost(2, 3), c(1, 4)
$\qquad$ + cost(2, 4), c(1, 5) + cost(2, 5)}
$\qquad$ = min{9 + 7, 7 + 9, 3 + 18, 2 + 15}
$\qquad$ = min{16, 16, 21, 17}
$\qquad$ = 16

A minimum-cost *s*-to-*t* path has a cost of 16. This path can be determined easily if we record the decision made at each state (vertex).

Let **d(i, j)** be the value of *l* (where *l* is a node) that minimizes **c(j, l) + cost(i + 1, l)** In figure,

d(3, 6) = 10
d(3, 7) = 10
d(3, 8) = 10
d(2, 2) = 7

$d(2, 3) = 6$
$d(2, 4) = 8$
$d(2, 5) = 8$
$d(1, 1) = 2.$

Let the minimum-cost path be

$$1, V_2, V_3, \ldots\ldots\ldots\ldots\ldots\ldots V_{k-1}, t$$
$$V_2 = d(1,1) = 2$$
$$V_3 = d(2,d(1,1)) = d(2,2) = 7$$
$$V_4 = d(3,d(2,d(1,1))) = d(3,7) = 10$$

So the solution minimum cost path is 1, 2, 7, 10, 12.

Backward Approach

The multistage graph problem can also be solved using the backward approach. Let **bp(i, j)** be a minimum-cost path from vertex **s** to a vertex **j** in *Vi*, Let **bcost(i, j)** be the cost of **bp(i, j)**. From the backward approach, we obtain

$$bcost(i, j) = \min\{bcost(i-1, l) + c(l, j)\}$$

$$l \in V_{i-1}'$$

$$\langle l, j \rangle \in E.$$

$$Since\, bcost(2, j) = c(1, j)\, if\, \langle 1, j \rangle \in E\, and\, bcost(2, j) = \infty\, if\, \langle 1, j \rangle \notin E,$$

where **k − 2 ≤ i ≤ k** for a **k**-stage graph.
First compute bcost for i = 3, then for i = 4, and so on.
The path is found by

$$1, V_1, V_2, \ldots\ldots, V_{k-1}, t$$
$$V_k = t$$
$$V_{k-1} = d(k, t)$$
$$V_i = d(i + 1, d(i + 2, d(k, t)))$$

Multistage-Graph Pseudo Code Corresponding to the Backward Approach

```
void BGraph(graph G, int k, int n, int p[])
// Same function as FGraph
{
        float bcost[MAXSIZE]; int d[MAXSIZE], r;
        bcost [1] = 0.0;
        for (int j=2; j <= n; j++) { // Compute bcost[j].
```

let r be such that <r,j> is an edge of
G and bcost[r] + c[r] [j] is minimum;
bcost[j]= bcost[r] + c[r] [j];
d [j] = r;
}
//Find a minimum-cost path.
p[l] = 1; p[k] = n;
for (j = k-l; j >= 2; j--) p[j]= d [p [j+1]] ;
}

SOLUTION USING THE BACKWARD APPROACH

$bcost(3, 6) = \min\{bcost(2,2)+c(2,6),bcost(2,3)+c(3,6)\}$
$= \min\{9+4, 7+2\}$
$= \min\{13,9\} = 9.$

$bcost(3, 7) = \min\{bcost(2, 2) + c(2, 7), bcost(2, 3) + c(3, 7), bcost(2, 5) + c(5, 7)\}$
$= \min\{9 + 2, 7 + 7, 2 + 11\}$
$= \min\{11, 14, 13\}$
$= 11$

$bcost(3, 8) = \min\{bcost(2, 2) + c(2, 8), bcost(2, 4) + c(4, 8), bcost(2, 5) + c(5, 8)\}$
$= \min\{9 + 1, 3 + 11, 2 + 8\}$
$= \min\{10, 14, 10\}$
$= 10$

$bcost(4, 9) = \min\{bcost(3, 6) + c(6, 9), bcost(3, 7) + c(7, 9)\}$
$= \min\{9 + 6, 11 + 4\}$
$= \min\{15, 15\}$
$= 15$

$bcost(4, 10) = \min\{bcost(3, 6) + c(6, 10), bcost(3, 7) + c(7, 10), bcost(3, 8)$
$+ c(8, 10)\}$
$= \min\{9 + 5, 11 + 3, 10 + 5\}$
$= \min\{14, 14, 15\}$
$= 14$

$bcost(4, 11) = \min\{bcost(3, 8) + c(8, 11)\}$
$= \min\{10 + 6\}$
$= \min\{16\}$
$= 16$

$bcost(5, 12) = \min\{bcost(4, 9) + c(9, 12), bcost(4, 10) + c(10, 12), bcost(4, 11)$
$+ c(11, 12)\}$
$= \min\{15 + 4, 14 + 2, 16 + 5\}$
$= \min\{19, 16, 21\}$
$= 16$

A minimum-cost *s*-to-*t* path has a cost of 16. This path can be determined easily if
we record the decision made at each state (vertex) as shown in Figure 3.2.

V_1 V_2 V_3 V_4 V_5

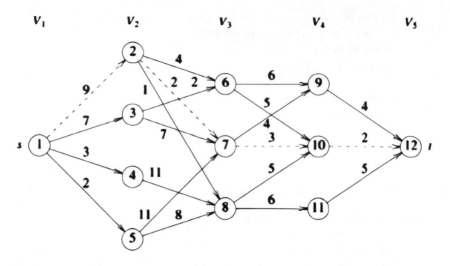

FIGURE 3.2   A minimum-cost *s*-to-*t* path is indicated by the broken edges

Let **d(i, j)** be the value of *l* (where *l* is a node) that minimizes **bcost(i–1, l)** + $o(l, j)$

In Figure 3.2,

d(3, 6) = 3
d(3, 7) = 2
d(3, 8) = 2
d(4, 9) = 6
d(4, 10) = 7
d(4, 11) = 8
d(5, 12) = 10

Let the minimum cost path be

$$1, V_2, V_3, \ldots\ldots\ldots\ldots\ldots\ldots, V_{k-1}, t$$
$$V_{k-1} = d(5, 12) = 10 = d(k, t)$$
$$V_3 = d(3 + 1, d(5, 12)) = d(4, 10) = 7$$
$$V_2 = d(3, d(4,10)) = d(3, 7) = 2$$

So the solution minimum cost path is 1, 2, 7, 10, 12.

3.3   ALL-PAIRS SHORTEST PATHS

The all-pairs shortest path problem involves finding the shortest path from each node in the graph to every other node in the graph. Let G = (V, E) be a directed graph with **i** vertices. Let **cost** be a cost of adjacency matrix for G such that cost(i, i) = 0, $1 \leq i \leq n$. Then cost(i, j) is the length (or cost) of edge <i, j> if $< i, j > \in E(G)$ and cost(i, j) = ∞

if $i \neq j$ and $<i, j> \notin E(G)$. The **all-pairs shortest path problem** is to determine a matrix **A** such that **A(i, j)** is the length of the shortest path from **i** to **j**.

- A **directed graph** is a graph with directions on the edges.
- A **labeled directed graph** is a directed graph with positive costs on the edges.
- A **path** in a graph G = (V, E) is a sequence of edges.
- The **length** of a path is the number of edges.
- The cost of a path is the sum of the costs of the edges.

PROCEDURE

- The **matrix A** can be obtained by solving **n** single-source problems using the algorithm ShortestPaths.
- Let us examine a shortest **i** to **j** path in G, $i \neq j$. This path originates at vertex **i**, goes through some intermediate vertices (possibly none), and terminates at vertex **j**.
- This path contains no cycles. If there is a cycle, then this can be deleted without increasing the path length (no cycle has a negative length).
- If **k** is an intermediate vertex on this shortest path, then the sub-paths from **i** to **k** and from **k** to **j** must be the shortest paths from **i** to **k** and **k** to **j**, respectively. Otherwise, the *i*-to-*j* path is not of minimum length. So, the **principle of optimality** holds. This alerts us to the prospect of using dynamic programming.
- If **k** is the intermediate vertex with the highest index, then the **i-to-k** path is a shortest **i-to-k** path in **G** going through no vertex with index greater than $k - 1$.
- Similarly, the **k-to-j** path is the shortest **k-to-j** path in **G** going through no vertex of index greater than $k - 1$.

Using $\Lambda^k(i, j)$ to represent the length of the shortest path from **i** to **j** going through no vertex of index greater than **k**, we obtain

$$A(i,j) = \min\left\{\min_{1\le k \le n}\left\{A^{k-1}(i,k) + A^{k-1}(k,j)\right\} cost(i,j)\right\}.$$

Clearly, $A^0(i,j) = cost(i,j), 1 \le i \le n, 1 \le j \le n.$

The shortest path from **i** to **j** going through no vertex higher than **k** either goes through vertex **k** or it does not. If it does,

$$A^k(i,j) = A^{k-1}(i,k) + A^{k-1}(k,j).$$

If it does not, then no intermediate vertex has an index greater than $k - 1$. Hence,

$$A^k(i,j) = A^{k-1}(i,j).$$

Combining, we get

$$A^k(i,j) = \min\left\{A^{k-1}(i,j), A^{k-1}(i,k) + A^{k-1}(k,j)\right\}, k \geq 1.$$

FUNCTION TO COMPUTE LENGTHS OF THE SHORTEST PATHS

```
void AllPaths(float cost[] [SIZE], float A[] [SIZE], int n)
// cost[1:n] [1:n] is the cost adjacency matrix of a graph with n vertices;
// A[i] [j] is the cost of a shortest path from vertex i to vertex j.
// cost [i] [i] = 0.0, for 1 ≤ i ≤ n.
{
        for (int i=1; i<=n; i++)
        for (int j=1; j<=n; j++)
        A[i] [j] = cost[i] [j];   // Copy cost into A.
        for (int k=1; k<=n; k++)
        for (i=1; i<=n; i++)
        for (int j=1; j<=n; j++)
        A[i] [j] = min(A[i] trr. A[i] [k]+A[k] [j]);
}
```

Example

$$A^{(0)} = \begin{bmatrix} 0 & \infty & 3 & \infty \\ 2 & 0 & \infty & \infty \\ \infty & 7 & 7 & 1 \\ 6 & \infty & \infty & 0 \end{bmatrix}$$

$A_{ij}^{(0)}$ equals

- the cost of the edge from i to j if $i \neq j$ and $\langle i,j \rangle \in E(G)$.
- ∞ if $i \neq j$ and $\langle i,j \rangle \notin E$.
- 0 if i = j.

$A^{(1)}$ = the length of the shortest path with intermediate vertices numbered no higher than 1.

$$A^{(1)} = \begin{bmatrix} 0 & \infty & 3 & \infty \\ 2 & & & \\ \infty & & & \\ 6 & & & \end{bmatrix}$$

Now apply the formula to the remaining elements.

$$A^k(i,j) = \min\left\{A^{k-1}(i,j), A^{k-1}(i,k) + A^{k-1}(k,j)\right\}, k \geq 1$$

$$A_{ij}^{(k)} = \min\left\{A_{ij}^{(k-1)}, A_{ik}^{(k-1)} + A_{kj}^{(k-1)}\right\}$$

$$A_{22}^{(1)} = \min\left\{A_{22}^{(0)}, A_{21}^{(0)} + A_{12}^{(0)}\right\}$$

$$A_{22}^{(1)} = \min\left\{0, 2 + \infty\right\} = 0$$

$$A_{23}^{(1)} = \min\left\{A_{23}^{(0)}, A_{21}^{(0)} + A_{13}^{(0)}\right\}$$

$$A_{23}^{(1)} = \min\left\{\infty, 2 + 3\right\} = 5$$

$$A_{24}^{(1)} = \min\left\{\infty, 2 + \infty\right\} = \infty$$

$$A_{32}^{(1)} = \min\left\{A_{32}^{(0)}, A_{31}^{(0)} + A_{12}^{(0)}\right\}$$

$$A_{32}^{(1)} = \min\left\{7, \infty + \infty\right\} = 7$$

$$A_{33}^{(1)} = \min\left\{A_{33}^{(0)}, A_{31}^{(0)} + A_{13}^{(0)}\right\}$$

$$A_{33}^{(1)} = \min\left\{0, \infty + 3\right\} = 0$$

$$A_{33}^{(1)} = \min\left\{1, \infty + \infty\right\} = 1$$

$$A_{42}^{(1)} = \min\left\{A_{42}^{(0)}, A_{41}^{(0)} + A_{12}^{(0)}\right\}$$

$$A_{42}^{(1)} = \min\left\{\infty, 6 + \infty\right\} = \infty$$

$$A_{43}^{(1)} = \min\left\{A_{43}^{(0)}, A_{41}^{(0)} + A_{13}^{(0)}\right\}$$

$$A_{43}^{(1)} = \min\left\{\infty, 6 + 3\right\} = 9$$

$$A_{44}^{(1)} = \min\left\{A_{44}^{(0)}, A_{41}^{(0)} + A_{14}^{(0)}\right\}$$

$$A_{44}^{(1)} = \min\left\{0, 6 + \infty\right\} = 0$$

So

$$A^{(1)} = \begin{bmatrix} 0 & \infty & 3 & \infty \\ 2 & 0 & 5 & \infty \\ \infty & 7 & 0 & 1 \\ 6 & \infty & 9 & 0 \end{bmatrix}.$$

$A^{(2)}$ = the length of the shortest path with intermediate vertices numbered no higher than 2.

$$A^{(1)} = \begin{bmatrix} & & \infty & \\ 2 & 0 & 5 & \infty \\ & & 7 & \\ & & \infty & \end{bmatrix}$$

Applying the following formula,

$$A_{ij}^{(k)} = \min\left\{A_{ij}^{(k-1)}, A_{ik}^{(k-1)} + A_{kj}^{(k-1)}\right\},$$

we get

$$A^{(2)} = \begin{bmatrix} 0 & \infty & 3 & \infty \\ 2 & 0 & 5 & \infty \\ 9 & 7 & 0 & 1 \\ 6 & \infty & 9 & 0 \end{bmatrix}.$$

Similarly,

$$A^{(3)} = \begin{bmatrix} & & 3 & \\ & & 5 & \\ 9 & 7 & 0 & 1 \\ & & 9 & \end{bmatrix}$$

Applying the formula, we get

$$A^{(3)} = \begin{bmatrix} 0 & 10 & 3 & 4 \\ 2 & 0 & 5 & 6 \\ 9 & 7 & 0 & 1 \\ 6 & 16 & 9 & 0 \end{bmatrix}.$$

For n vertices, we have to find up to $A^{(n)}$.

$$A^{(4)} = \begin{bmatrix} & & & 4 \\ & & & 6 \\ & & & 1 \\ 6 & 16 & 9 & 0 \end{bmatrix}$$

The resultant matrix is

$$A^{(4)} = \begin{bmatrix} 0 & 10 & 3 & 4 \\ 2 & 0 & 5 & 6 \\ 7 & 7 & 0 & 1 \\ 6 & 16 & 9 & 0 \end{bmatrix}.$$

Note: The running time of Floyd's algorithm is $\theta\left(n^3\right)$.

3.4 OPTIMAL BINARY SEARCH TREES

An optimal binary search tree (OBST) is one special kind of advanced tree. It focuses on how to reduce the cost of the search of the binary search tree (BST). Given a fixed set of identifiers, different BSTs have been created for the same identifiers set to have different performance characteristics. A BST is one of the most important data structures in computer science. One of its principal applications is to implement a dictionary, a set of elements with the operations of searching, insertion, and deletion.

It needs three tables to record probabilities, cost, and root. The identifier having a higher probability of appearing should be placed nearer to the root, and the identifier having a lower probability of appearing should be placed away from the root. The BST created with such kind of arrangement is called an OBST.

The tree of Figure 3.3a, in the worst case, requires four comparisons to find an identifier, whereas the tree of Figure 3.3b requires only three.

For example, in the case of tree (a), it takes 1, 2, 2, 3, and 4 comparisons, respectively, to find the identifiers **for**, **do**, **while**, **int**, and **if**. Thus, the average number of comparisons is $\dfrac{1+2+2+3+4}{5}=\dfrac{12}{5}$. In the case of tree (b), it takes 1, 2, 2, 3, and 3 comparisons, respectively, to find the identifiers **for**, **do**, **int**, **if**, and **while**. Thus, the average number of comparisons is $\dfrac{1+2+2+3+3}{5}=\dfrac{11}{5}$. It requires only $\dfrac{11}{5}$ comparisons. This calculation assumes that each identifier is searched for with equal probability and that no unsuccessful searches (i.e., searches for identifiers not in the tree) are made.

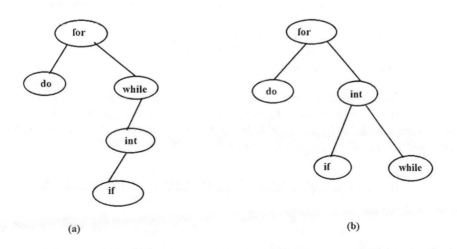

(a)   (b)

FIGURE 3.3   Two possible BSTs

GENERAL DEFINITION

Let $\{a_1, a_2, \ldots a_n\}$ be a set of identifiers with $a_1 < a_2 < \ldots \ldots < a_n$. Let **p(i)** be the probability with which we search for $\mathbf{a_i}$. Let **q(i)** be the probability that the identifier x being searched for is such that $\mathbf{a_i < x < a_{i+1}}$. $0 \leq i \leq n$.

Then $\displaystyle\sum_{i=1}^{n} p(i)$ is the probability of a successful search.

$\displaystyle\sum_{i=0}^{n} q(i)$ is the probability of an unsuccessful search

Thus, the OBST is constructed with the optimum cost for

$$\sum_{i=1}^{n} p(i) + \sum_{i=0}^{n} q(i) = 1.$$

COST FUNCTION

To obtain the cost function for BSTs, it is useful to add a fictitious node in place of every empty subtree in the search tree. Such nodes are called as external nodes. All other nodes are internal nodes.

If a binary search tree represents **n** identifiers, then there will be exactly **n** internal nodes and **n + 1** (fictitious) external nodes. Every internal node represents a point where a successful search may terminate. Every external node represents a point where an unsuccessful search may terminate.

Successful search: If a successful search terminates at an internal node at level '*l*', then *l* iterations are needed. Hence, the expected cost contribution from the internal node for $\mathbf{a_i}$ is **p(i) * level(a_i)**.

Unsuccessful search: An unsuccessful search terminates at the external nodes. The identifiers not in the BST can be partitioned into **n + 1** equivalence classes $\mathbf{E_i}$, $0 \leq i \leq n$.

Since there will be **n + 1** external nodes in a binary tree with **n** identifiers,

$\mathbf{E_0}$ contains all identifiers **x** such that $\mathbf{x < a_i}$.
$\mathbf{E_i}$ contains all identifiers **x** such that $\mathbf{E_i < x < a_{i+1}}$, **1 ≤ i ≤ n**.
$\mathbf{E_n}$ contains all identifiers **x** such that $\mathbf{x > a_i}$.

All identifiers in the same class $\mathbf{E_i}$, the search terminates at the same external node. For identifiers in different $\mathbf{E_i}$, the search terminates at different external nodes. If the failure node for $\mathbf{E_i}$, is at level *l*, then only *l* − 1 iterations are made. Hence, the cost contribution of this node is **q(i) * (level (E_i − 1))**.

$$\text{Expected cost of } \mathbf{a} = \frac{\text{Cost for successful search+binary search tree}}{\text{Cost for unsuccessful search}}$$

We define an optimal binary search tree for the identifier set $\{a_1, a_2, \ldots a_n\}$ to be a BST for which $\displaystyle\sum_{1 \leq i \leq n} p(i) \times level(a_i) + \sum_{0 \leq i \leq n} q(i) \times (level(E_i) - 1)$ is the minimum.

OBST CONSTRUCTION ALGORITHM

Definition: An OBST is a binary search tree with a minimum weighted path length; that is,

$$\sum_{1\leq i\leq n} p(i)\times level(a_i)+ \sum_{0\leq i\leq n} q(i)\times\left(level(E_i)-1\right) \text{ should be the minimum.}$$

To apply dynamic programming to the problem of obtaining an OBST, we need to view the construction of such a tree as the result of a sequence of decisions and then observe that the principle of optimality holds when applied to the problem state resulting from a decision. A possible approach to this would be to make a decision as to which of the a_i's should be assigned to the root node of the tree.

The expected cost of the left subtree is given by

$$cost(l)= \sum_{1\leq i\leq k} p(i)\times level(a_i)+ \sum_{0\leq i\leq k} q(i)\times\left(level(E_i)-1\right).$$

The expected cost of the right subtree is given by

$$cost(r)= \sum_{k\leq i\leq n} p(i)\times level(a_i)+ \sum_{k<i\leq n} q(i)\times\left(level(E_i)-1\right).$$

An OBST with root a_k is shown in Figure 3.4,

GENERAL FORMULA

$$\mathbf{w(l, j) = p(j) + q(j) + w(i, j - 1)}$$

$$c(i,j) = w(i,j)+ \min\left\{c(i,k-1)+c(k,j)\right\}$$

$$i \leq k \leq j$$

$$\mathbf{r(i, j) = \text{value of k that minimizes } c(i, j).}$$

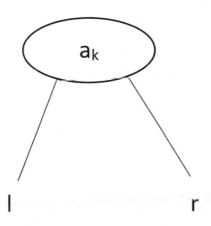

FIGURE 3.4   OBST with root a_k

FINDING A MINIMUM-COST BST ALGORITHM

void OBST(float p[], float q[], int n)

// *Given n distinct identifiers al < a2 < . . . < an and probabilities p[i], $1 \leq i \leq n$,*
   and

//*q [i], $0 \leq i \leq n$, this algorithm computes the cost c[i] [j] of optimal binary*
   search trees

// t_{ij} *for identifiers $a_{i+1}, \ldots, a_j$. It also computes r [i] [j], the root of t_{ij}. w [i] [j]*
   is the //weight of t_{ij} .

```
{
        float c[SIZE][SIZE], w[SIZE] [SIZE]; int r[SIZE] [SIZE];
        for (int i=0; i <= n–1; i++) { // Initialize.
        w [i][i]= q [i]; r [i][i] = 0; c[i][i]=0.0;
        //Optimal trees with one node
        w [i] [i+1] = q [i] + q [i+1] + p [i+1] ;
        r[iJ [i+1] = i + 1;
        c [i] [i+1] = q [i] + q [i+1] + p [i+1] ;
        }
        w[n] [n] = q[n]; r[n] [n] = 0; c[n] [n] = 0.0;
        for (int m=2; m<= n; m++) // Find optimal trees with m nodes.
        for (i=0; i <= n-m; i++) {
        int j = i + m,
        w[i][j] = w[i][j–1] + p [j ] + q[j];
        int k = Find(c,r,i,j);
        // A value of 1 in the range
        // r[iJ[j–1] <= 1 <= r[i+1][j] that
        // minimizes {c[i] [1–1]+c[l] [j]};
        c [i][j] = w [i] [j] + c [i] [k–1] + c [k] [j] ;
        r [i][j] = k;
        }
        cout << c [0] [n]<< " " << w [0] [n] <<" " << r [0][n];
}
int Find(float c[] [SIZE], int r[] [SIZE],int i,int j)
{
        float min=INFTY; int 1;
        for (int m=r[i] [j–1]; m<=r[i+1] [j]; m++)
        if ((c [i] [m-1] +c [m] [j]) < min) {
        min = c[i] [m-1]+c[m][j]; 1 = m;
        }
        return(l);
}
```

Example

Obtain the OBST for the following:

n = 4, (a_1, a_2, a_3, a_4) = (do, if, int, while) p (1:4) = (3, 3, 1, 1) q (0:4) = (2, 3, 1, 1, 1)

Solution
Initially $0 \le i \le n$

$w(i, i) = q[i]$
$w(0, 0) = q[0] = 2$
$w(1, 1) = q[1] = 3$
$w(2, 2) = q[2] = 1$
$w(3, 3) = q[3] = 1$
$w(4, 4) = q[4] = 1$
$c(i, i) = 0$
$c(0, 0) = 0$
$c(1, 1) = 0$
$c(2, 2) = 0$
$c(3, 3) = 0$
$c(4, 4) = 0$

$r(i, i) = 0$
$r(0, 0) = 0$
$r(1, 1) = 0$
$r(2, 2) = 0$
$r(3, 3) = 0$
$r(4, 4) = 0$

Formula

$$w(i, j) = p(j) + q(j) + w(i, j - 1)$$

$$c(i,j) = w(i,j) + \min\{c(i,k-1) + c(k,j)\}$$

$$i \le k \le j$$

$$r(i, j) = \text{value of k that minimizes } c(i, j)$$

TO COMPUTE FOR W(I, I + 1), C(I, I + 1), AND R(I, I + 1)

$$w(0, 1) = p(1) + q(1) + w(0, 0)$$

$$w(0, 1) = 3 + 3 + 2 = 8$$

$$c(0,1) = w(0,1) + \min\{c(i,k-1) + c(k,j)\}$$

$$0 \le k \le 1$$

$$c(0,1) = w(0,1) + \min\{c(0,0) + c(1,1)\}$$

$$c(0,1) = 8 + \min\{0 + 0\} = 8$$

r (0, 1) = 1

w (1, 2) = p(2) + q(2) + w(1, 1)

w (1, 2) = 3 + 1 + 3 = 7

$$c(1,2) = w(1,2) + \min\{c(i,k-1) + c(k,j)\}$$

$$1 \le k \le 2$$

$$c(1,2) = w(1,2) + \min\{c(1,1) + c(2,2)\}$$

$$c(1,2) = 7 + \min\{0,0\} = 7$$

r (1, 2) = 2

w (2, 3) = p(3) + q(3) + w(2, 2)

w (2, 3) = 1 + 1 + 1 = 3

$$c(2,3) = w(2,3) + \min\{c(i,k-1) + c(k,j)\}$$

$$2 \le k \le 3$$

$$c(2,3) = w(2,3) + \min\{c(2,2) + c(3,3)\}$$

$$c(2,3) = 3 + \min\{0,0\} = 3$$

r (2, 3) = 3

w (3, 4) = p(4) + q(4) + w(3, 3)

w (3, 4) = 1 + 1 + 1 = 3

$$c(3,4) = w(3,4) + \min\{c(i,k-1) + c(k,j)\}$$

$$3 \le k \le 4$$

$$c(3,4) = w(3,4) + \min\{c(3,3) + c(4,4)\}$$

$$c(3,4) = 3 + \min\{0,0\} = 3$$

r (3, 4) = 4

To Compute for w(i, i + 2), c(i, i + 2), and r(i, i + 2)

w (0, 2) = p(2) + q(2) + w(0, 1)

w (0, 2) = 3 + 1 + 8 = 12

$$c(0,2) = w(0,2) + \min\{c(i,k-1) + c(k,j)\}$$

$$0 \le k \le 2$$

$$c(0,2) = w(0,20 + \min\{c(0,0) + c(0,1) + c(2,2)\}$$

$$c(0,2) = 12 + \min\{0 + 7,8 + 0\} = 12 + 7 = 19$$

r (0, 2) = 1

w (1, 3) = p(3) + q(3) + w(1, 2)

w (1, 3) = 1 + 1 + 7 = 9

$$c(1,3) = w(1,3) + \min\{c(i,k-1) + c(k,j)\}$$

$$1 \le k \le 3$$

$$c(1,3) = w(1,3) + \min\{(c(1,1) + c(2,3) + c(1,2) + c(3,3)\}$$

$$c(1,3) = 9 + \min\{0 + 3,7 + 0\} = 9 + 3 = 12$$

r (1, 3) = 2

w (2, 4) = p(4) + q(4) + w(2, 3)

w (2, 4) = 1 + 1 + 3 = 5

$$c(2,4) = w(2,4) + \min\{c(i,k-1) + c(k,j)\}$$

$$2 \le k \le 4$$

$$c(2,4) = w(2,4) + \min\{c(2,2) + c(3,4), c(2,3) + c(4,4)\}$$

$$c(2,4) = 5 + \min\{0 + 3,3 + 0\} = 5 + 3 = 8$$

r (2, 4) = 3

TO COMPUTE FOR W(I, I + 3), C(I, I + 3), AND R(I, I + 3)

w (0, 3) = p(3) + q(3) + w(0, 2)

w (0, 3) = 1 + 1 + 12 = 14

$$c(0,3) = w(0,3) + \min\{c(i,k-1) + c(k,j)\}$$

$$0 \le k \le 3$$

$$c(0,3) = w(0,3) + \min\{c(0,0) + c(1,3), c(0,1) + c(2,3), c(0,2) + c(3,3)\}$$

$$c(0,3) = 14 + \min\{0 + 12, 8 + 3, 19 + 0\} = 14 + 11 = 25$$

r (0, 3) = 2

w (1, 4) = p(4) + q(4) + w(1, 3)

w (1, 4) = 1 + 1 + 9 = 11

$$c(1,4) = w(1,4) + \min\{c(i, k-1) + c(k, j)\}$$

$$1 \le k \le 4$$

$$c(1,4) = w(1,4) + \min\{c(1,1) + c(2,4), c(1,2) + c(3,4), c(1,3) + c(4,4)\}$$

$$c(1,4) = 11 + \min\{0 + 8, 7 + 3, 12 = 0\} = 11 + 8 = 19$$

r (1, 4) = 2

To Compute for w(i, i + 4), c(i, i + 4), and r(i, i + 4)

w (0, 4) = p(4) + q(4) + w(0, 3)

w (0, 4) = 1 + 1 + 14 = 16

$$c(0,4) = w(0,4) + \min\{c(i, k-1) + c(k, j)\}$$

$$0 \le k \le 4$$

$$c(0,4) = w(0,4) + \min\{c(0,0) + c(1,4), c(0,1) + c(2,4), c(0,2)$$
$$+ c(3,4), c(0,3) + c(4,4)\}$$

$$c(0,4) = 16 + \min\{0 + 19, 8 + 8, 19 + 3, 25 + 0\} = 16 + 16 = 32$$

r (0, 4) = 2

The computation of c(0, 4), w(0, 4), and r(0, 4) is shown in Figure 3.5.
Thus, c(0, 4) = 32 is the minimum cost of a BST for (a_1, a_2, a_3, a_4).

The root of tree is R(0, 4) = 2.   (a_2)

a_2 is 'if'; now assume i = 0, j = 4, and k = 2.

Left child $\Rightarrow r_{i,k}-_1 = r_{01} = 1$ (a_1)

a_1 is 'do' Right child $\Rightarrow r_{k,j} = r_{24} = 3$ (a_3)

a_3 is 'int'; now assume that i = 2, j = 4, and k = 3.

$$\text{Right child of } a_3 \Rightarrow r_{k,j} = r_{42} = 4 \tag{a_4}$$

a_4 is 'while'.

Now the OBST is given as shown in Figure 3.6.

TIME COMPLEXITY OR TIME ANALYSIS

c(i, j) are computed for j − i = 1, 2, ... , n, if j − i = m, there are (n − m) + 1 c(i, j)'s to compute. Each c(i, j) requires to compute **m** quantities. Hence, each c(i, j) can be computed in O(m) time. The total time f or all c(i, j)'s with j − i = m is

$$O(n - m + 1)(m) = O(nm - m^2 + m) = O(nm - m^2).$$

The total time to evaluate all the c(i, j)'s and r(i, j)'s is

$$\sum_{1 \leq m \leq n} \left(nm - m^2\right) = n\sum m - \sum m^2$$

$$= \frac{n(m)(m+1)}{2} - \frac{m(m+1)(2m+1)}{6}$$

$$= O(n^3).$$

	0	1	2	3	4
0	$w_{00} = 2$ $c_{00} = 0$ $r_{00} = 0$	$w_{11} = 3$ $c_{11} = 0$ $r_{11} = 0$	$w_{22} = 1$ $c_{22} = 0$ $r_{22} = 0$	$w_{33} = 1$ $c_{33} = 0$ $r_{33} = 0$	$w_{44} = 1$ $c_{44} = 0$ $r_{44} = 0$
1	$w_{01} = 8$ $c_{01} = 8$ $r_{01} = 1$	$w_{12} = 7$ $c_{12} = 7$ $r_{12} = 2$	$w_{23} = 3$ $c_{23} = 3$ $r_{23} = 3$	$w_{34} = 3$ $c_{34} = 3$ $r_{34} = 4$	
2	$w_{02} = 12$ $c_{02} = 19$ $r_{02} = 1$	$w_{13} = 9$ $c_{13} = 12$ $r_{13} = 2$	$w_{24} = 5$ $c_{24} = 8$ $r_{24} = 3$		
3	$w_{03} = 14$ $c_{03} = 25$ $r_{03} = 2$	$w_{14} = 11$ $c_{14} = 19$ $r_{14} = 2$			
4	$w_{04} = 16$ $c_{04} = 32$ $r_{04} = 2$				

FIGURE 3.5   Computation of c(0, 4), w(0, 4), and r(0, 4)

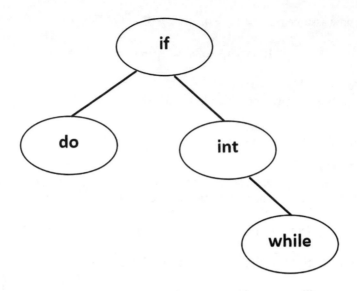

FIGURE 3.6   OBST

We know that

$$c(i,j) = w(i,j) + \min\{c(i,k-1) + c(k,j)\}$$

$$i \le k \le j.$$

Time complexity can be reduced to $O(n^2)$ using D. E. Knuth's result, which limits the search to the range

$$r(i,j-1) \le k < r(i+1,j) \text{ instead of } w(i,j) + \min\{c(i,k-1) + c(k,j)\}, 1 \le k \le j.$$

The tree t_{0n} can be constructed from the values r(i, j) in O(n) time.

Brute Force Method
- It is an exhaustive search.
- The number of BSTs that can be constructed with n keys is equal to the nth Catalan number.

$$c(n) = \binom{2n}{n} \frac{1}{n+1} \text{ for } n > 0 \quad c(0) = 1$$

For example:
    When key = 2, that is, n = 2,

$$c(n) = \binom{4}{2}\frac{1}{3}$$

$$= 4C_2 \frac{1}{3}$$

$$= \frac{4 \times 3}{2 \times 1} \cdot \frac{1}{3} = 2$$

The possible BST is 2. That is, then the tree is $a_1 < a_2$ as shown in Figure 3.7.

If we have three nodes, where $a_1 < a_2 < a_3$, then the number of the possible binary search tree is five as shown in Figure 3.8.

$$c(n) = \binom{6}{3}\frac{1}{4}$$

$$= \frac{6 \times 5 \times 4}{3 \times 2 \times 1} \cdot \frac{1}{4} = 5$$

Example Problem

The possible binary search tree for the identifiers $(a_1, a_2, a_3) = $ (do, if, while). Consider equal probabilities $p(i) = q(i) = 1/7$ for all i.

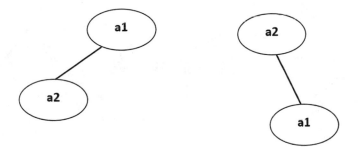

FIGURE 3.7 BST $a_1 < a_2$

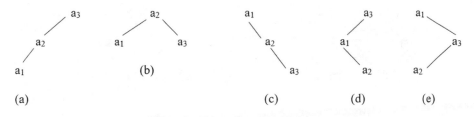

FIGURE 3.8 BST $a_1 < a_2 < a_3$

Solution

We have three nodes. So the number of possible $BST = c(n) = \binom{6}{3}\frac{1}{4}$

$$= \frac{6 \times 5 \times 4}{3 \times 2 \times 1} \cdot \frac{1}{4} = 5.$$

The trees are with equal probability $p(i) = q(i) = \frac{1}{7}$.

Let us find the optimal binary search tree out of the five trees shown in Figure 3.9.

$$cost(tree\,a) = \sum_{i=1}^{n} p(i)*level(a_i) + \sum_{i=0}^{n} q(i)*(level(E_i)-1)$$

$$= \frac{1}{7}[1+2+3] + \frac{1}{7}[(4-1)+(4-1)+(3-1)+(2-1)]$$

$$= \frac{6}{7} + \frac{1}{7}[3+3+2+1]$$

$$= \frac{6}{7} + \frac{9}{7} = \frac{15}{7}$$

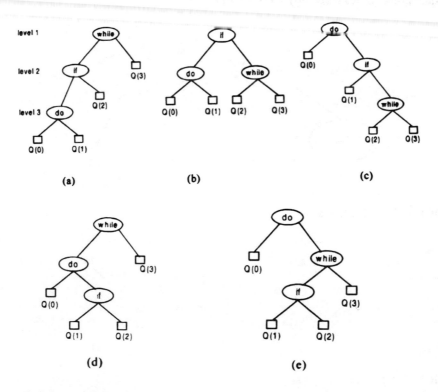

FIGURE 3.9   OBST example

$$cost(tree\,b) = \frac{1}{7}[1+2+2] + \frac{1}{7}[(3-1)+(3-1)+(3-1)+(3-1)]$$

$$= \frac{5}{7} + \frac{1}{7}[2+2+2+2]$$

$$= \frac{5}{7} + \frac{8}{7} = \frac{13}{7}$$

$$cost(tree\,c) = \frac{1}{7}[1+2+3] + \frac{1}{7}[(4-1)+(4-1)+(3-1)+(2-1)]$$

$$= \frac{6}{7} + \frac{1}{7}[3+3+2+1]$$

$$= \frac{6}{7} + \frac{9}{7} = \frac{15}{7}$$

$$cost(tree\,d) = \frac{1}{7}[1+2+3] + \frac{1}{7}[(4-1)+(4-1)+(3-1)+(2-1)]$$

$$= \frac{6}{7} + \frac{1}{7}[3+3+2+1]$$

$$= \frac{6}{7} + \frac{9}{7} = \frac{15}{7}$$

$$cost(tree\,e) = \frac{1}{7}[1+2+3] + \frac{1}{7}[(4-1)+(4-1)+(3-1)+(2-1)]$$

$$= \frac{6}{7} + \frac{1}{7}[3+3+2+1]$$

$$= \frac{6}{7} + \frac{9}{7} = \frac{15}{7}$$

Tree (b) is optimal.

Example: 2
Consider p(1) = 0.5, p(2) = 0.1, p(3) = 0.05, q(0) = 0.15, q(1) = 0.1, q(2) = 0.05, and q(3) = 0.05 for the identifiers (a_1, a_2, a_3) = (do, if, while).

Solution

$$cost(tree\,a) = \sum_{i=1}^{n} p(i) * level(ai) + \sum_{i=0}^{n} q(i) * (level(Ei) - 1)$$

$$= p(1) * level(a1) + p(2) * level(a2) + p(3) * level(a3) + q(0) * (level(Ei) - 1)$$

$$+ q(1) * (level(Ei) - 1) + q(2) * (level(Ei) - 1) + q(3) * (level(Ei) - 1)$$

$$= 0.5 * 3 + 0.1 * 2 + 0.05 * 1 + 0.15 * 3 + 0.1 * 3 + 0.05 * 2 + 0.05 * 1$$

$$= 1.5 + 0.2 + 0.05 + 0.45 + 0.3 + 0.1 + 0.05$$

$$= 2.65$$

$$\cos t(\text{tree b}) = 0.5*2 + 0.1*1 + 0.05*2 + 0.15*2 + 0.1*2 + 0.05*2 + 0.05*2$$
$$= 1.0 + 0.1 + 0.1 + 0.3 + 0.2 + 0.1 + 0.1$$
$$= 1.9$$

$$\cos t(\text{tree c}) = 0.5*1 + 0.1*2 + 0.05*3 + 0.15*1 + 0.1*1 + 0.05*3 + 0.05*3$$
$$= 0.5 + 0.2 + 0.15 + 0.15 + 0.1 + 0.15 + 0.15$$
$$= 1.5$$

$$\cos t(\text{tree d}) = 0.05*1 + 0.5*2 + 0.1*3 + 0.15*2 + 0.1*3 + 0.05*3 + 0.05*1$$
$$= 0.05 + 1.0 + 0.3 + 0.3 + 0.3 + 0.15 + 0.05$$
$$= 1.35 + 0.8$$
$$= 2.15$$

$$\cos t(\text{tree e}) = 0.5*1 + 0.05*2 + 0.1*3 + 0.15*1 + 0.1*3 + 0.05*3 + 0.05*2$$
$$= 0.5 + 0.1 + 0.3 + 0.15 + 0.3 + 0.15 + 0.1$$
$$= 0.9 + 0.7 = 1.6$$

Tree **c** has a smaller value. Hence, tree **c** is optimal BST.

EXERCISE

Compute w(i, j), r(i, j), and c(i, j), $0 \le i < j \le 4$ for the identifier set (a_1, a_2, a_3, a_4) = (cout, float, if, while) with p(1) = 1/20, p(2) = 1/5, p(3) = 1/10, p(4) = 1/20, q(0) = 1/5, q(1) = 1/10, q(2) = 1/5, q(3) = 1/20, and q(4) = 1/20. Using the r(i, j)'s construct, the OBST.

3.5   0/1 KNAPSACK

Given **n** objects and a knapsack where object i has a weight W_i and the knapsack has a capacity 'm'. If a fraction x_i, $0 \le x_i \le 1$ of object 'i' placed into the knapsack, then a profit $P_i x_i$ is earned. The objective is to obtain filling the knapsack to maximize the total profit.

SOLUTION

The solution to the knapsack problem can be obtained by making a sequence of decisions on the variables $x_1, x_2, x_3, \ldots \ldots, x_n$, and the decision on variable x_i involves determining which of the values 0 or 1 is to be assigned to it.

DOMINANCE RULES OR PURGING RULES

If S^{i+1} contains two pairs (P_j, W_j) and (P_k, W_k) with the property that $P_j \le P_k$ and $W_j \ge W_k$, then the pair (P_j, W_j) is discarded. This is known as a dominance rule or a purging rule.

PROCEDURE TO COMPUTE THE SOLUTION FOR 0/1 KNAPSACK

1. Set $S^0 = \{(0, 0)\}$.
2. Compute S_1^i using the formula
   $S_1^i = \{(P, W) \mid (P + P_{i+1}, W + W_{i+1}) \in S^i\}$.
3. Compute S^{i+1} by merging S^i and S_1^i.
4. After merging, if $S_i + 1$ contains two pairs (P_j, W_j) and (P_k, W_k) such that $P_j \leq P_k$ and $W_j \geq W_k$, then the pair (P_j, W_j) can be discarded. This type of discarding pair is called a purging rule or a dominance rule.
5. Then discard the pair containing weight greater than the given weight (m).
6. Repeat the procedure up to 'n' times or choose the pair containing maximum profit as the optimal solution.

EXAMPLE

Consider the knapsack instance n = 3 $(W_1, W_2, W_3) = (2, 3, 4)$, $(P_1, P_2, P_3) = (1, 2, 5)$, and m = 6.

$$P_i = 1 \quad 2 \quad 5$$
$$W_i = 2 \quad 3 \quad 4$$

SOLUTION

Step 1: Initialize S^0

$$i = 0, \qquad S^0 = \{(0, 0)\}$$
$$p, w$$

Find S_1^0

Add the first object. $S_1^0 = \{(P + P_{i+1}, W + W_{i+1})\}$

So, $S_1^0 = \{(0 + 1, 0 + 2)\} = \{(1, 2)\}$.

Step 2: Find S^1

$S^1 = S^0 + S_1^0 \Rightarrow$ That is, merge $(S^0 \text{ and } S_1^0)$.

$S^1 = \{(0, 0) \ (1, 2)\}$

$i = 1, S_1^1 = \{(P + P_{i+1}, W + W_{i+1})\}$ Add the next object.

$= \{(0 + 2, 0 + 3) \ (1 + 2, 2 + 3)\}$

$= \{(2, 3)(3, 5)\}$

Step 3: Find S^2

$S^2 = S^1 + S_1^1 \Rightarrow$ That is, merge (S^1 and S_1^1).

$S^2 = \{(0, 0)(1, 2)(2, 3)(3, 5)\}$ Check whether it satisfies purging rule.

$S_1^2 = \{(P + P_{i+1}, W + W_{i+1})\}$ Add the next object.

$\qquad = \{(0 + 5, 0 + 4)(1 + 5, 2 + 4)(2 + 5, 3 + 4)(3 + 5, 5 + 4)\}$

$\qquad = \{(5, 4)(6, 6)(7, 7)(8, 9)\}$

Step 4: Find S^3

$S^3 = S^2 + S_1^2 \Rightarrow$ That is, merge (S^2 and S_1^2).

$S^3 = \{(0, 0)(1, 2)(2, 3)(3, 5)(5, 4)(6, 6)(7, 7)(8, 9)\}$

Check whether it satisfies the purging rule. If it satisfies the condition $P_j \leq P_k \, and \, W_j \geq W_k$, then discard ($P_j$, W_j).

Here the pair (3, 5) has been eliminated from S^3. Because it satisfies $P_j \leq P_k \, and \, W_j \geq W_k$, 3 < 5 and 5 > 4. Then discard the points (7, 7) and (8, 9). Here, m = 6, 7 > 6, and 9 > 6, so discard.

Now $S^3 = \{(0, 0)(1, 2)(2, 3)(5, 4)(6, 6)\}$.

Step 5: Computation of Xi's

Now the maximum profit obtained is (6, 6).

1. Check if (6, 6) is in S^2 or not. If it is available, put $X_3 = 0$; otherwise, put $X_3 = 1$

   $(6,6) \notin S^2$. So $X_3 = 1$. {For Si, $X_{i+1} = 0$ or $X_{i+1} = 1$.}

   $(6 - P_3, 6 - W_3) \Rightarrow (6 - 5, 6 - 4) \Rightarrow (1, 2)$

2. Check if (1, 2) is in S^1 or not.

   $(1,2) \in S^1$. So $X_2 = 0$.

3. Check if (1, 2) is in S^0 or not.

   $(1,2) \notin S^0$. So $X_1 = 1$.

   $(1 - P_1, 2 - W_1) \Rightarrow (1 - 1, 2 - 2) \Rightarrow (0, 0)$.

Here an optimal solution is $(X_1, X_2, X_3) = (1, 0, 1)$.

INFORMAL KNAPSACK ALGORITHM

```
DKP(P,W,n,m)
{
S⁰= {(0, 0)}
for (i=1;i<=n−1;i++)
{
S₁ⁱ⁻¹ ={(P,W) |(P-Pᵢ, W-Wᵢ)∈ Sⁱ⁻¹ and W≤m};
Si=MergePurge(Sⁱ⁻¹,S₁ ⁱ⁻¹);
}
```

(Px, Wx) = last pair in S^{n-1}
(Py, Wy) = $(P^1 + P_n, W^1 + W_n)$, where W^1 is the largest W in any pair in S^{n-1}
   such that
W + Wn $\leq$ m;
If (Px>Py)
X_n = 0;
else,
X_n = 1;
Trace back for $(X_{n-1}, \ldots\ldots\ldots, X_1)$;
}

3.6   THE TRAVELING SALESPERSON PROBLEM

DEFINITION

Let $G = (V, E)$ be a directed graph with edge costs C_{ij}. The variable C_{ij} is defined
such that $C_{ij} > 0$ for all i and j and $C_{ij} = \infty$ if $<i, j> \notin E$.

Let $|V| = n$ and assume $n > 1$. A **tour** of G is a directed simple cycle that includes
every vertex in V. The cost of a tour is the sum of the cost of the edges on the tour.
The *traveling salesperson problem* is to find a tour of minimum cost. Edge $<i, j>$ is
assigned a cost equal to the distance from site i to site j.

APPLICATIONS

1. Routing a postal van to pick up mail from mailboxes located at n different
   sites
   • Suppose we have to route a postal van to pick up mail from mailboxes
     located at n different sites.
   • An $n + 1$ vertex graph can be used to represent the situation.
   • One vertex represents the post office from which the postal van starts
     and to which it must return. Edge (i, j) is assigned a cost equal to the
     distance from site i to site j.
   • The route taken by the postal van is a tour, and we have to find a tour of
     minimum length.
2. Planning a robot arm to tighten the nuts on different positions
   • Suppose we wish to use a robot arm to tighten the nuts on some piece of
     machinery on an assembly line.
   • The arm will start from its initial position (which is over the first nut
     to be tightened), successively move to each of the remaining nuts, and
     return to the initial position.
   • The path of the arm is clearly a tour on a graph in which vertices repre-
     sent the nuts.
   • A minimum-cost tour will minimize the time needed for the arm to com-
     plete its task
3. Planning production in which n different commodities are manufactured on
   the same sets of machines

- From a production environment several commodities are manufactured on the same set of machines. The manufacture proceeds in cycles. In each production cycle, **n** *different* commodities are produced.
- When the machines are changed from the production of commodity **i** to commodity **j**, a changeover cost C_{ij} is incurred.
- It is desired to find a sequence in which to manufacture these commodities. This sequence should minimize the sum of change over costs
- Since the manufacture proceeds cyclically, it is necessary to include the cost of starting the next cycle. This is just the changeover cost from the last to the first commodity. Hence, this problem can be regarded as a traveling salesperson problem on an **n** vertex graph with edge cost C_{ij} being the changeover cost from commodity **i** to commodity **j**.

THE TRAVELING SALESPERSON AND THE PRINCIPLE OF OPTIMALITY

A tour is to be a simple path that starts and ends at vertex **1**. Every tour consists of an edge **(1, k)** for some **k** and a path from vertex **k** to vertex **1**. The path from vertex **k** to vertex **1** goes through each vertex in **V − {1, k}** exactly once. It is easy to see that if the tour is optimal, then the path from **k** to **1** must be the shortest **k**-to-**1** path going through all vertices in **V − {1, k}**. Hence, the principle of optimality holds.

DYNAMIC PROGRAMMING FORMULATION

Let **g(i, S)** be the length of the shortest path starting at vertex **i**, going through all vertices in **S**, and terminating at vertex **1**.

The function **g(1, V − {1})** is the length of an optimal salesperson tour. From the principle of optimality, it follows that

$$g\left(1, V - \{1\}\right) = min_{2 \le k \le n} \left\{ c_{1k} + g\left(k, V - \{1, k\}\right) \right\}.$$

Generalizing this equation,

$$g\left(i, S\right) = min_{j \in s} \left\{ C_{ij} + g\left(j, S - \{j\}\right) \right\}.$$

PROCEDURE

$$g\left(i, \varnothing\right) = C_{i1} \qquad\qquad 2 \le i \le n$$
$$g\left(i, S\right) \text{ for all } |S| = 1 \qquad 2 \le i \le n$$
$$g\left(i, S\right) \text{ for all } |S| = 2 \qquad 2 \le i \le n$$

.

.

$$g\left(i, S\right) \text{ for all } |S| = n - 2 \qquad 2 \le i \le n$$
$$g\left(i, S\right) \text{ for all } |S| = n - 1 \qquad S = \{V - \{1\}\}$$

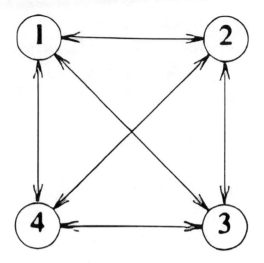

FIGURE 3.10   Directed graph

Consider the following directed graph shown in Figure 3.10 and the edge length given by the matrix.

$$\begin{bmatrix} 0 & 10 & 15 & 20 \\ 5 & 0 & 9 & 10 \\ 6 & 13 & 0 & 12 \\ 8 & 8 & 9 & 0 \end{bmatrix}$$

$$|S| = 0$$
$$g(i,\varnothing) = C_{i1}$$
$$g(2,\varnothing) = C_{21} = 5$$
$$g(3,\varnothing) = C_{31} = 6$$
$$g(4,\varnothing) = C_{41} = 8$$
$$|S| = 1$$
$$i = 2 \text{ to } n$$

Here, n = 4 (4 vertices), so i varies from 2 to 4.

$$g(i,S) = min_{j \in S} \left\{ C_{ij} + g(j, S - \{j\}) \right\}$$
$$\text{If } i = 2,$$
$$g(2,\{3\}) = min_{j \in 3} \left\{ C_{23} + g(3,\{3\} - \{3\}) \right\}$$
$$g(2,\{3\}) = min \left\{ C_{23} + g(3,\varnothing) \right\}$$

$$g(2,\{3\}) = \min\{9+6\} = 15$$
$$g(2,\{4\}) = min_{j\in4}\{C_{24} + g(4,\{4\}-\{4\})\}$$
$$g(2,\{4\}) = \min\{C_{24} + g(4,\varnothing)\}$$
$$g(2,\{4\}) = \min\{10+8\} = 18.$$

If i = 3,
$$g(3,\{2\}) = min_{j\in2}\{C_{32} + g(2,\{2\}-\{2\})\}$$
$$g(3,\{2\}) = \min\{C_{32} + g(2,\varnothing)\}$$
$$g(3,\{2\}) = \min\{13+5\} = 18$$
$$g(3,\{4\}) = min_{j\in4}\{C_{34} + g(4,\{4\}-\{4\})\}$$
$$g(3,\{4\}) = \min\{C_{34} + g(4,\varnothing)\}$$
$$g(3,\{4\}) = \min\{12+8\} = 20.$$

If i = 4,
$$g(4,\{2\}) = min_{j\in2}\{C_{42} + g(2,\{2\}-\{2\})\}$$
$$g(4,\{2\}) = \min\{C_{42} + g(2,\varnothing)\}$$
$$g(4,\{2\}) = \min\{8+5\} = 13$$
$$g(4,\{3\}) = min_{j\in3}\{C_{43} + g(3,\{3\}-\{3\})\}$$
$$g(4,\{3\}) = \min\{C_{43} + g(3,\varnothing)\}$$
$$g(4,\{3\}) = \min\{9+6\} = 15$$
$$|S| = 2.$$

i = 2 to n
$$g(2,\{3,4\}) = min_{j\in3,4}\{C_{23} + g(3,\{3,4\}-\{3\}), C_{24} + g(4,\{3,4\}-\{4\})\}$$
$$g(2,\{3,4\}) = \min\{9 + g(3,4), 10 + g(4,3)\}$$
$$g(2,\{3,4\}) = \min\{9+20, 10+15\} = \min\{29,25\} = 25$$

i = 3
$$g(3,\{2,4\}) = min_{j\in2,4}\{C_{32} + g(2,\{2,4\}-\{2\}), C_{34} + g(4,\{2,4\}-\{4\})\}$$
$$g(3,\{2,4\}) = \min\{13 + g(2,4), 12 + g(4,2)\}$$
$$g(3,\{2,4\}) = \min\{13+18, 12+13\} = \min\{31,25\} = 25$$

$$i = 4$$

$$g(4,\{2,3\}) = min_{j \in 2,3}\{C_{42} + g(2,\{2,3\}-\{2\}), C_{43} + g(3,\{2,3\}-\{3\})\}$$

$$g(4,\{2,3\}) = min\{C_{42} + g(2,3), C_{43} + g(3,2)\}$$

$$g(4,\{2,3\}) = min\{8+15,9+18\} = min\{23,27\} = 23$$

$$|S| = 3$$

$$i = 2 \text{ to } n$$

$$g(1,\{2,3,4\}) = min_{j \in 2,3,4}\{C_{12} + g(2,\{2,3,4\}-\{2\}),$$

$$C_{13} + g(3,\{2,3,4\}-\{3\}), C_{14} + g(4,\{2,3,4\}-\{4\})\}$$

$$g(1,\{2,3,4\}) = min\{\{C_{12} + g(2,\{3,4\}), C_{13} + g(3,\{2,4\}), C_{14} + g(4,\{2,3\})\}$$

$$g(1,\{2,3,4\}) = min\{10+25,15+25,20+23\} = min\{35,40,43\} = 35$$

To find the order of nodes to be reached:

$$g(1,\{2,3,4\}) = min\{35,40,43\} = 35$$

Here, 35 is the minimum, so select node **2**
   So select the edge from **1** *to* **2**.

$$g(2,\{3,4\}) = min\{29,25\} = 25$$

Here, 25 is the minimum. So select node **4**.
   So select the edge from **2** *to* **4**.

Then, consider $g(4,\{3\}) = min\{9+16\} = 15$.

So select node **3**. *So select the edge from* **4** *to* **3**. Then return to node **1**.
   So the tour path is 1, 2, 4, 3, 1.

Analysis

$$Time\,Complexity = O(n^2 2^n)$$

$$Space\,Complexity = O(n2^n)$$

Note
Edge from 1 to 2 = 10
Edge from 2 to 4 = 10
Edge from 4 to 3 = 9
Edge from 3 to 1 = 6 Total cost = 35 (Tour of minimum cost)

4 Backtracking

4.1 BACKTRACKING: THE GENERAL METHOD

Backtracking is a more intelligent variation of the algorithmic approach. The principal idea is to construct solutions one component at a time and evaluate such partially constructed candidates as follows. If a partially constructed solution can be developed further without violating the problem's constraints, it is done by taking the first remaining legitimate option for the next component. If there is no legitimate option for the next component, no alternatives for *any* remaining component need to be considered. In this case, the algorithm backtracks to replace the last component of the partially constructed solution with its next option.

It is convenient to implement this kind of processing by constructing a tree of choices being made, called the ***state-space tree***. Its root represents an initial state before the search for a solution begins. The nodes of the first level in the tree represent the choices made for the first component of a solution, the nodes of the second level represent the choices for the second component, and so on. A node in a state-space tree is said to be ***promising*** if it corresponds to a partially constructed solution that may still lead to a complete solution; otherwise, it is called ***nonpromising***. Leaves represent either nonpromising dead ends or complete solutions found by the algorithm. In the majority of cases, a state-space tree for a backtracking algorithm is constructed in the manner of a depth-first search. If the current node is promising, its child is generated by adding the first remaining legitimate option for the next component of a solution, and the processing moves to this child. If the current node turns out to be nonpromising, the algorithm backtracks to the node's parent to consider the next possible option for its last component; if there is no such option, it backtracks one more level up the tree and so on. Finally, if the algorithm reaches a complete solution to the problem, it either stops (if just one solution is required) or continues searching for other possible solutions.

For fundamental principles of algorithm design, backtracking represents one of the most general techniques. Many problems that deal with searching for a set of solutions or that ask for an optimal solution satisfying some constraints can be solved using the backtracking formulation. Find a solution by trying one of several choices. If the choice proves incorrect, the computation backtracks or restarts at the point of choice and tries another choice. It is often convenient to maintain choice points and alternate choices using recursion.

Basic Idea of Backtracking

- Backtracking is one of the most general techniques for searching a set of solutions or for searching an optimal solution.
- The name 'backtrack' was first coined by D. H. Lehmer in the 1950s.

DOI: 10.1201/9781003355403-4

- For some problems, the only way to solve is to check all possibilities.
- To apply the backtracking method, the solution must be expressed as an n-tuple $(x_1, x_2, \ldots, x_n)$, where each $x_i \in S_i$, with S_i being a finite set.

CONSTRAINTS

- Solutions must satisfy a set of constraints.
- Constraints can be divided into two categories:
  - Explicit
  - Implicit

Explicit Constraints: Explicit constraints are rules that restrict each x_i to take on values only from a given set.

Example

$x_i \geq 0$ or $S_i = \{$all nonnegative real numbers$\}$
$x_i = 0$ or 1 or $S_i = \{0,1\}$

All tuples satisfying the explicit constraints define a possible solution space for I (I = problem instance).

Implicit Constraints: The implicit constraints are rules that determine which of the tuples in the solution space of I satisfy the criterion function. Thus, implicit constraints describe the way in which the x_i must relate to each other.

Example: 8-Queens Problem

Recursive Backtracking Algorithm

```
void Backtrack(int k)
// This is a schema that describes the backtracking process using recursion.
// On entering the first k – 1 values x[1], x[2], . . . , x[k – 1] of the solution vector
// x[1:n] have been assigned. x [] and n are global.
{
        for (each x[k] such that x[k] ∈ T(x[1], . . . , x [k–1])
        {
                if (Bₖ(x[1], x[2], , x[k])) {
                if (x [1], x [2], , x [k] is a path to an answer node)
                output x [1: k];
                if (k < n) Backtrack(k+1);
        }
    }
}
```

General Backtracking Method

```
void IBacktrack(int n)
// This is a schema that describes the backtracking process.
//All solutions are generated in x[1:n] and printed
// as soon as they are determined.
{
        int k=1;
        while (k) {
                if (there remains an untried x[k] such that
                        x [k] is in T(x[1], x[2], . . . , x[k–1]) and
                        B_k(x [1], . . . , x [k]) is true) {
                        if (x[1], . . . , x[k] is a path to an answer node) output
                        x[1:k];
                        k++; // Consider the next set.
                }
                else k--; // Backtrack to the previous set.
        }
}
```

TERMINOLOGY

Solution space: Tuples that satisfy the explicit constraints define a solution space. The solution space can be organized into a tree.

Problem state: Problem state is each node in the depth-first search tree.

State space: State space is the set of all paths from the root node to the other nodes.

Solution states: Solution states are the problem states s for which the path from the root node to s defines a tuple in the solution space.

- In variable-tuple size formulation tree, all nodes are solution states.
- In fixed-tuple size formulation tree, only the leaf nodes are solution states.
- Partitioned into disjoint sub-solution spaces at each internal node.

Answer states: Answer states are those solution states s for which the path from root node to s defines a tuple that is a member of the set of solutions

- These states satisfy implicit constraints.

State-space trees: State-space trees are the tree organization of the solution space.

Static trees: Static trees are ones for which tree organizations are independent of the problem instance being solved.

- Fixed-tuple size formulation
- The tree organization is independent of the problem instance being solved.

Dynamic trees: Dynamic trees are ones for which the organization is dependent on the problem instance.

Live nodes: Live nodes are the generated nodes for which all the children have not yet been generated.

E-Node: E-node is a live node whose children are currently being generated and expanded.

Dead nodes: Dead nodes are the generated nodes that are not to be expanded any further or for which all their children have been generated.

- All the children of a dead node are already generated.
- Live nodes are killed using a bounding function to make them dead nodes.
- These states satisfy the implicit constraints.

Bounding function: The bounding function is a function used to kill live nodes without generating all their children.

Backtracking: Backtracking is depth-first node generation with bounding functions.

Branch and bound: Branch and bound is a state generation method in which E-node remains E-node until it is dead.

Breadth-first search: Branch and bound with each new node placed in a queue. The front of the queue becomes the new E-node.

Depth search (D-search): New nodes are placed into a stack. The last node added is the first to be explored.

4.2   THE 8-QUEENS PROBLEM

Consider an **n × n** chessboard and try to find all ways to place *n* nonattacking queens.

Consider the 4-queens problem; here, $(X_1, \ldots \ldots, X_n)$ represents a solution in which X_i is the column of the *i*th row where the *i*th queen is placed. All X_i's will be distinct since no two queens can be placed in the same column.

If we imagine a chessboard's squares being numbered as the indices of the two-dimensional array **a [1: n] [1: n]**, then every element on the same diagonal that runs from the upper left to the lower right has the same 'row–column' value.

8-QUEENS PROBLEM

It is a classic combinatorial problem that places eight queens on an 8 × 8 chessboard so that no two queens attack each other. That is, no two queens should be placed on same row, column, or diagonal. The 8 × 8 chessboard is numbered from 1 to 8 (rows and columns). The queens are numbered from 1 to 8 as shown in Figure 4.1. Since each queen must be on a different row, each queen must be placed on row i.

Explicit criteria: S = {1, 2, 3, . . . , 8}. Define set 'S,' which has 8 values since it is an 8 × 8 matrix.

Implicit criteria: No two queens must attack each other. No two x_i's can be same.

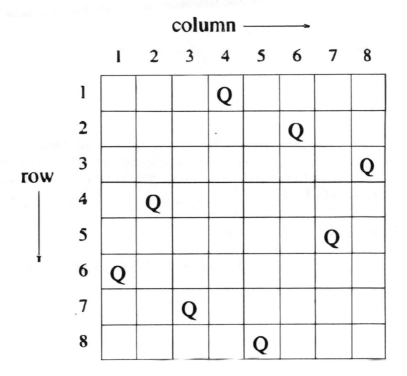

FIGURE 4.1   Board of the 8-queens problem

- For example, consider the queen at **a[4][2]**. The squares that are diagonal to this queen are **a[3][1]**, **a[5][3]**, **a[6][4]**, **a[7][5]**, and **a[8][6]**.
- All these squares have a row–column value of 2.
- Every element on the same diagonal that goes from the upper right to the lower left has the same 'row + column' value.

Suppose two queens are placed at positions (i, j) and (k, l). Then they are on the same diagonal only if $i - j = k - l$ or $i + j = k + l$.

$$\text{The first equation implies } j \quad l = i - k.$$
$$\text{The second equation implies } j - l = k - i.$$

Algorithm: Can a New Queen Be Placed?

bool Place(int k, int i)

// Returns true if a queen can be placed in kth row and ith column. Otherwise it //returns false. x[] is a global array whose first (k − 1) values have been set.

// abs(r) returns the absolute value of r.

```
{
        for (int j=l; j < k; j++)
        if ((x[j] == i)                    // Two in the same column
                || (abs(x[j]-i)== abs(j-k))) // or in the same diagonal
                return(false);
        return(true);
}
```

Place(k,i) is a function that returns a Boolean value that is true if **kth** queen can be placed in column **i**. It tests both whether **i** is distinct from all previous values $x[1], \ldots, x[k-1]$ and whether there is no other queen on the same diagonal. Its computing time is $O\ (k-1)$.

All Solutions to the N-Queens Problem

```
void NQueens(int k, int n)
// Using backtracking, this procedure prints all possible placements of n
    queens
//on an n × n chessboard so that they are nonattacking.
{
        for (int i=l; i<=n; i++) {
            if (Place(k, i)) {
                    x[k] = i;
                    if (k==n) { for (int j=l;j<=n;j++)
                    cout <<x[j] <<' '; cout << endl;}
                    else NQueens(k+l, n);
            }
        }
}
```

For an 8 × 8 chessboard, there are $\binom{64}{8}$ possible ways to place eight pieces as shown in Figure 4.2. We can use Estimate to estimate the number of nodes that will be generated by N-queens.

Note that the assumptions that are needed for **Estimate** do hold for N-queens. The bounding function is static. No change is made to the function as the search proceeds. In addition, all nodes on the same level of the state-space tree have the same degree. In Figure 4.2, we see five 8 × 8 chessboards that were created using **Estimate**. The total number of nodes in the 8-queens state-space tree is

$$1+\sum_{j=0}^{7}\left[\prod_{i=0}^{j}(8-i)\right] = 69.281.$$

So the estimated number of unbounded nodes is only about 2.34% of the total number of nodes in the 8-queens state-space tree.

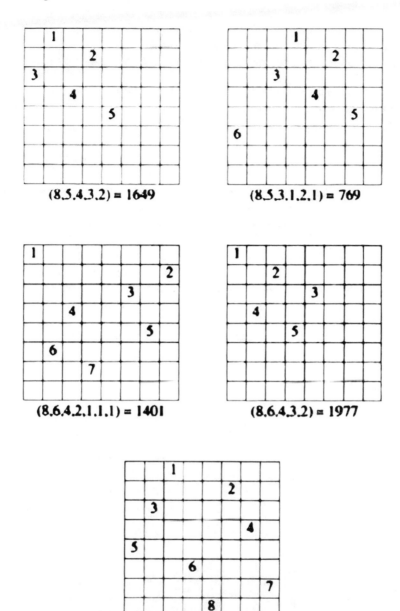

(8.5.4.3.2) = 1649

FIGURE 4.2 Solving the 8-queens problem

4-QUEENS PROBLEM

It is a classic combinatorial problem that places four queens on a 4 × 4 chessboard so that no two queens attack each other. That is, no two queens should be placed on the same row, column, or diagonal. The 4 × 4 chessboard is numbered from 1 to 4 (rows and columns). The queens are numbered from 1 to 4.

Since each queen must be on different row, each queen must be placed on row **i**.

Tree Organization of 4-Queens Solution Space (Nodes are Numbered as in a Depth-first Search)

A possible tree organization for the case **n = 4** is in Figure 4.3. This is called a permutation tree. The edges are labeled by possible values of x_i. Edges from level 1 to level 2 nodes specify the values for x_1. Thus, the leftmost sub-tree contains all solutions with $x_1 = 1$.

Its leftmost sub-tree contains all solutions with $x_1 = 1$ and $x_2 = 2$ and so on. Edges from level **i** to level **i + 1** are labeled with the values of x_i. The solution space is defined by all paths from the root node to a leaf node. There are 4! = 24 leaf nodes in the tree.

> **State space:** Each node in the tree defines the problem state. All paths from the root to the nodes are called the state space of the problem.
>
> **Answer states:** *Answer states* are those solution states **s** for which the path from the root to 's' defines a tuple that is a member of the set of solutions of the problem.
>
> **State-space tree:** The tree organization of the solution space is referred to as the *state-space tree*.
>
> **Live node:** A node that has been generated and whose children have not yet all been generated is called a *live node*.

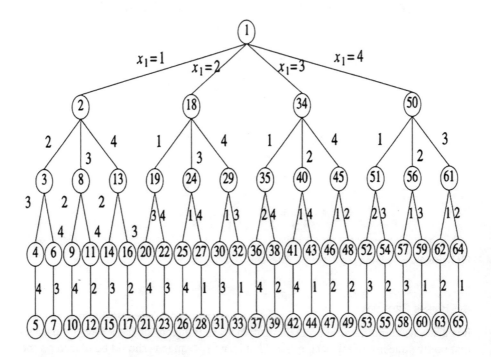

FIGURE 4.3 Tree organization of the 4-queens solution space

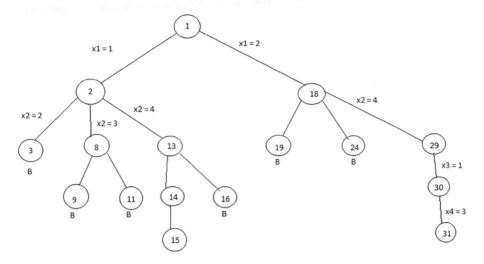

FIGURE 4.4   Example of the 4-queens problem

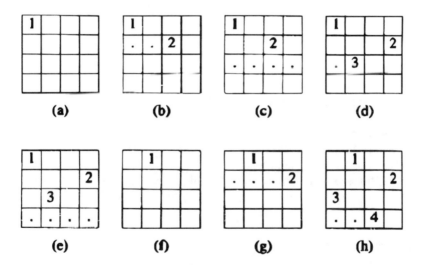

FIGURE 4.5   Solving the 4-queens problem

E-node: The live node whose children are currently being generated is called the *E-node*.

Dead node: A *dead node* is a generated node that is not to be expanded further or whose children have not all been generated.

Figure 4.4 shows how backtracking works on the 4-queens problem, and the solution is shown in Figure 4.5.

- The first node is a live node or E-node.
- The second node is also generated from the first node with path = 1. This corresponds to placing a queen in the first column.

- Node 2 is an E-node.
- Now, node 3 is generated from the second node with a path length of 2 (Path = 1, 2). So this corresponds to placing the second queen in the second column.
- Now, node 3 is killed; it is assumed that node 2 is a dead node.
- Backtrack to node 2; node 2 is an E-node, so node 8 is generated with a path = [1, 3].
- Now, node 9 is generated with a path = [1, 3, 2].
- So, node 9 is killed and is assumed to be a dead node.
- Backtrack to node 8; from this, node 11 is generated with a path = [1, 3, 4].
- So, node 11 is also killed; it is the dead node. Finally, node 8 is also killed because we can't place a third queen with reference to the position of the second queen.
- Now backtrack to node 2.
- Now from node 2, node 13 is generated with a path = [1, 4].
- From node 13, node 14 is generated with a path = [1, 4, 2].
- From node 14, node 15 is generated with a path = [1, 4, 2, 3].
- So node 15 is killed; it is called a dead node. Now backtrack to node 14. So kill node 14 also because no other possibilities to place the fourth queen from node 14.
- Backtrack to node 13, and node 16 is generated with a path = [1, 4, 3].
- So kill node 16 and kill node 13, and the path we arrive at is wrong, so finally node 2 is also killed.
- Backtrack to node 1, which generates node 18 with a path − [?]
- Now node 19 is generated from node 18 with a path [2, 1].
- So node 19 is killed. Backtrack to node 18.
- Now node 24 is generated with a path = [2, 3].
- So node 24 is killed. Backtrack to node 18.
- Now node 29 is generated with a path = [2, 4].
- Node 30 is generated from node 29 with a path = [2, 4, 1].
- Now node 31 is generated from node 30 with a path = [2, 4, 1, 3].

N-QUEENS PROBLEM

Consider an **n × n** chessboard and try to find all ways to place **n** nonattacking queens. Here $(X_1, \ldots\ldots, X_n)$ represents a solution in which X_i is the column of the **i**th row where the **i**th queen is placed. All X_is will be distinct since no two queens can be placed in the same column.

If we imagine a chessboard's squares being numbered as the indices of the two-dimensional array **a [l: n] [l: n]**, then every element on the same diagonal that runs from the upper left to the lower right has the same 'row–column' value.

4.3   SUM OF SUBSETS

Some positive integers will be given. Another integer 'm' is also given. We have to find the subset which must give the value 'm' when all the elements of subset are added.

Given distinct positive numbers (usually called weights) denoted by W_i, $1 \leq i \leq n$, and 'm'. Problem is to find all combinations of these numbers whose sums are m. This is called the *sum of subsets* problem.

Example: n = 4, w = {11, 13, 24, 7}, m = 31

To find all possible subsets that produce the sum = 31.

Each solution will have k-tuples. In this case, we have two tuples:

1. Variable-sized
2. Fixed-sized

VARIABLE-SIZED TUPLE/INDEX METHOD

For the earlier example,

m = 31, then {13 + 11 + 7} = 31 index = {1, 2, 4} also {24 + 7} = 31 index = {3, 4}

EXPLANATION

The tree of Figure 4.6 corresponds to the variable tuple size formulation. Here, start from node 1. Initially, x varies from 1 to 4. x_1 = 1, 2, 3, 4. From x_1, other variations are x_2 = **2, 3, 4**, and from x_2 = **2**, other variations are x_3 = **3, 4**. From x_3, next x_4 = 4. And here, only the increasing order of tree construction is used.

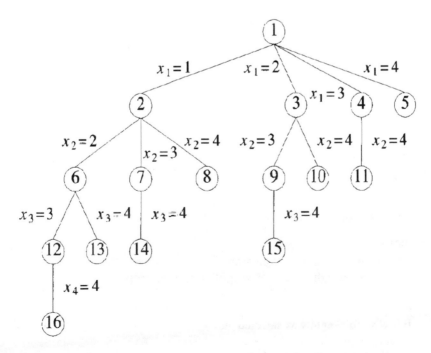

FIGURE 4.6   Nodes are numbered as in a breadth-first search

Given m = 31:

- Node 1 is the root node, which is an E-node.
- It generates node 2, it is an E-node, and it generates node 6.
- After generating node 2, [x_1 = 1]. So w_1 = **11**. So w = 11.
- After generating node 6, [x_1 = 1, x_2 = 2], so w_1 = **11** and w_2 = **13**. So w = 24.
- Node 6 is an E-node; it generates node 12.
- Now x_3 = 3, x_1 = 1, x_2 = 2. So w_1 = **11**, w_3 = **24**. So w = 48.
- It exceeds the maximum limit m = 31. So backtrack to node 6 and generate node 13 [x_1 = 1, x_2 = 2, x_3 = 4]. So w_1 = **11**, w_2 = **13**, w_4 = **7**.

Disadvantage: Backtracking will find only the first solution it finds. It will not see other solutions.

Fixed-Sized Tuple/Binary Method

In this case, the element X_i of the solution vector is either one or zero depending on whether the weight W_i is included. For a node at level **i**, the left child corresponds to $X_i = 1$ and the right to $X_i = 0$.

For example, n = 4, m = 31, w = [11, 13, 24, 7]

Solution 1: [1, 2, 4] = [1, 1, 0, 1]

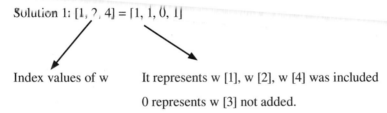

Index values of w   It represents w [1], w [2], w [4] was included

0 represents w [3] not added.

Solution 2: [3, 4] = [0, 0, 1, 1]

The tree is called static tree; the left child should have a value of 1, and the right child must have value 0. For any fixed-sized representation, the state space will have ($2^{n+1} - 1$) number of nodes. An example of the binary method is shown in Figure 4.7. For example, n = 4; that is, $2^{4+1} - 1 = 31$.

Method of Construction

Conditions
- **If selected weight > m, then stop.**
- **If (selected weight + sum of weight) < m, then stop.**

Explanation
- Initially, no weight was selected and level = 1; the remaining weights = all object weights (i.e., 11 + 13 + 24 + 7 = 55).
- So a node contains **[0, 1, 55]**

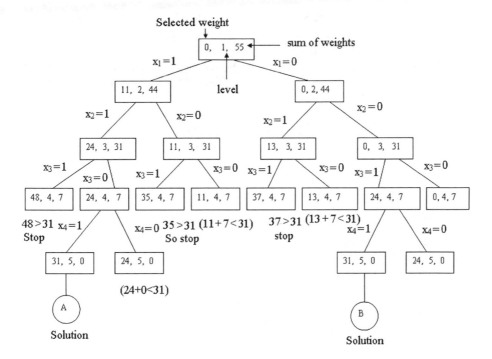

FIGURE 4.7 Example for binary method

- Always-left child has a value = 1.
- Always-right child has value = 0.
- Left child ($x_1 = 1$), Right child ($x_1 = 0$)
- Consider the left child ($x_1 = 1$); it means that weight $w_1 = 11$ was added, and now the sum of weights (55 − 11 = 44) is level = 2, so **[11, 2, 44]** is the second node.
- Consider the right child ($x_1 = 0$); it means w_1 was not selected. So the selected weight = 0, level = 2; now the sum of weights = 44. So **[0, 2, 44]** is the node.
- Finish the left child first and the right child next.
- Now **[11, 2, 44]** is selected from this two children, with the left child $x_2 = 1$ and right child $x_2 = 0$.
- When $x_2 = 1$, it means that w_2 was selected [$w_2 = 13$]. So '11 + 13' was selected, and the total sum of weight = 44 − 13 = 31. So the node becomes **[24, 3, 31]**.
- When $x_2 = 0$, now the right child starts to expand. We didn't add $w_2 = 13$. So the selected weight remains '11' only. We reduced the sum of weights to 31. So it only remains in the right child; now the node is **[11, 3, 31]**.
- From the node **[24, 3, 31]**, $x_3 = 1$ and $x_3 = 0$. If $x_3 = 1$ means $w_3 = 24$ is also added. Now the node is **[48, 4, 7]**. But 48 > 31, so stop its generation. If $x_3 = 0$ means $w_3 = 24$ was not selected. So the recently selected weight is only 24. The node becomes **[24, 4, 7]**.

- Now from the node [24, 4, 7], $x_4 = 1$ means $w_4 = 7$ was selected, so the total weight = 24 + 7 = 31, level = 5, and the sum of weights = 7 – 7 = 0. So the node becomes [31, 5, 0].
- From the node [24, 4, 7], $x_4 = 0$ means $w_4 = 7$ was not selected, so the weight selected = 24, level = 5, and the sum of weights = 7 – 7 = 0. So the node becomes [24, 5, 0].
- From [11, 3, 31] $x_3 = 1$ or $x_3 = 0$. If $x_3 = 1$ means $w_3 = 24$ was selected, but 11 + 24 = 35. 35 > 31. So stop, and it creates the node [35, 4, 7].
- From [11, 3, 31], if $x_3 = 0$ means $w_3 = 24$ was not selected. So the selected weight = 11 and level = 4, and the sum of weight = 7. It creates the node [11, 4, 7].

ALGORITHM

void SumOfSub(float s, int k, float r)

// Find all subsets of w[1:n] that sum to m. The values of x[j], 1<=j<k, have //

already been determined. $s = \sum_{j=1}^{k-1} w[j]^* x[j]$ and $r = \sum_{j=k}^{n} w[j]$

// The w[j]'s are in nondecreasing order.

// It is assumed that w [1]≤ m and $\sum_{i=1}^{n} w[i] \geq m$

```
{
// Generate left child. Note that s+w[k] <= m because B_{k-1} is true.
x[k] = 1;
if (s+w[k] == m) { // Subset found
for (int j=1; j<=k; j++) cout << x[j] <<' ';
cout <<endl;
}
// There is no recursive call here as w[j] > 0, 1 <= j <= n.
else if (s+w[k]+w[k+1] <= m)
SumOfSub(s+w[k], k+1, r-w[k]);
// Generate right child and evaluate B_k.
if ((s+r-w[k] >= m) && (s+w[k+1] <= m)) {
x[k] = 0;
SumOfSub(s, k+1, r-w[k]);
}
}
```

BOUNDING FUNCTIONS

A simple choice for the bounding functions is $B_k(X_1, \ldots, X_k)$ = true iff.

$$\sum_{i=1}^{k} W_i X_i + \sum_{i=k+1}^{n} W_i \geq m$$

Clearly $X_1, \ldots, X_k$ cannot lead to an answer node if this condition is not satisfied. The bounding functions can be strengthened if we assume the W_k's are initially in a nondecreasing order. In this case, $X_1, \ldots, X_k$ cannot lead to an answer node if

$$\sum_{i=1}^{k} w_i x_i + w_{k+1} > m.$$

The bounding functions we use are therefore

$$B_k\left(x_1 \ldots \ldots x_k\right) = true\, iff \sum_{i=1}^{k} w_i x_i + \sum_{i=k+1}^{n} w_i \geq m \ \ and \ \ \sum_{i=1}^{k} w_i x_i + w_{k+1} \leq m.$$

4.4 GRAPH COLORING

Let **G** be a graph and **m** be a given positive integer. M-colorability decision says whether the nodes of **G** can be colored in such a way that no two adjacent nodes have the same color and only **m** colors are used. This is termed the ***m-colorability decision*** problem. Note: If **d** is the degree of the given graph, then it can be colored with **d + 1** colors.

The ***m-colorability optimization*** problem asks for the smallest integer **m** for which graph G can be colored. This integer is referred to as the ***chromatic number*** of the graph.

For example, the graph can be colored with three colors, 1, 2, and 3, as shown in Figure 4.8. The color of each node is indicated next to it. It can also be seen that three colors are needed to color this graph and hence this graph's chromatic number is 3.

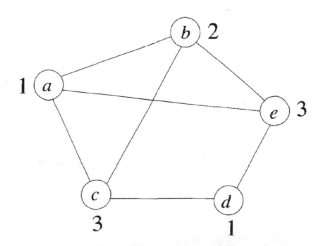

FIGURE 4.8   Graph coloring

PLANAR GRAPH

A graph is said to be *planar* iff it can be drawn in a plane in such a way that no two edges cross each other.

A famous special case of the m-colorability decision problem is the 4-color problem for planar graphs. This turns out to be a problem for which graphs are very useful, because a map can easily be transformed into a graph. Each region of the map becomes a node, and if two regions are adjacent, then the corresponding nodes are joined by an edge. Figure 4.9 shows a map with five regions and its corresponding graph. This map requires four colors.

Given any map, the regions can be colored in such a way that no two adjacent regions have the same color.

ALGORITHM

- Represent a graph by its adjacency matrix G[1:n, 1:n], where **G[i, j] = 1** if (i, j) is an edge of 'G' and **G[i, j] = 0** if (i, j) is not an edge of G.
- The colors are represented by the integers **1, 2, ..., m**, and the solutions are given by the n-tuple **(x₁, x₂, ... , xₙ)**, where x_i is the color of node **i**.
- Finding all m-colorings of the graph.

```
void mColoring(int k)
//This program was formed using the recursive backtracking schema.
// The graph is represented by its Boolean adjacency matrix G[1:n] [1:n].
// All assignments of 1, 2, . . . , m to the vertices of the graph such that adjacent
    vertices
//are assigned distinct integers are printed. k is the index of the next vertex to
    color.
{
        do { // Generate all legal assignments for x[k].
        NextValue(k); // Assign to x[k] a legal color.
```

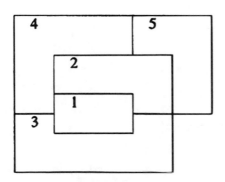 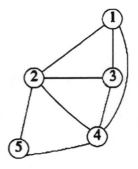

FIGURE 4.9　Planar graph

```
        if (!x[k]) break; // No new color possible
        if (k == n) { //At most m colors have been used to color the n vertices.
        for (int i=1; i<=n; i++) cout << x[i] << ' ';
        cout <<endl;
        }
        else mColoring(k+l);
        } while (1) ;
}
```

- Generating the next color

```
void NextValue(int k)
// x[1], . . . , x[k–1] have been assigned integer values in the range [1,m]
// such that adjacent vertices have distinct integers. A value for x[k] is
// determined in the range[0,m]. x[k] is assigned the next-highest-numbered
    color
// while maintaining distinctness from the adjacent vertices of vertex k.
//If no such color exists, then x[k] is zero.
{
        do {
        x[k] = (x[k]+1) % (m+1); // Next highest color
        if (!x[k]) return; // All colors have been used.
        for (int j-1; j<=n; j++) { // Check if this color is
                                    // distinct from adjacent colors.
        if (G[k] [j] // If (k, j) is an edge
            && (x[k]== x[j]))      // and if adj. vertices
                                    // have the same color
            break;
            }
            if (j == n+1) return; // New color found
        } while (1); // Otherwise try to find another color.
}
```

Example 1
  m = 3

METHOD OF CONSTRUCTION

- First we color node 1; we are given three colors. We can color node 1 with one of the three colors (1/2/3).
- So we have three children from root node in the tree.

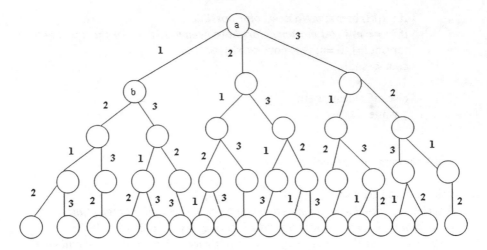

FIGURE 4.10   Solution tree

- If we select color 1 for node 1, then node 2 can be colored with either 2 or 3 so that we have two children from node 2.
- Now assume node 2 colored with color 2.
- Then node 3 can be colored with either color 1 or 3
- Again assume we selected color 1 for node 3. Then node 4 can be colored with 2 or 3.
- Now backtrack to node 3; if it was colored with color 3, then node 4 can be colored with color 2 alone only because node 1 and node 3 were colored with color 1 and color 3. So node 4 can be colored with color 2 only.
- Now backtrack to node 2; if it is colored with color 3, then node 3 can be colored with color 1 or 2.
- Now we assume node 3 is colored with color 1; then node 4 can be colored with color 2 or 3, and likewise, if node 3 is colored with color 2, then node 4 can be colored with color 3 only.
- Likewise, continue for the full solution tree as shown in Figure 4.10.

4.5   HAMILTONIAN CYCLES

Let $G = (V, E)$ be a connected graph with n vertices. A Hamiltonian cycle (suggested by Sir William Hamilton) is a round-trip path along n edges of G that visits every vertex once and returns to its starting position.

In other words, if a Hamiltonian cycle begins at some vertex $V_1 \in G$ and the vertices of G are visited in the order $V_1, V_2, \ldots, V_{n+1}$, then the edges $(Vi, Vi + 1)$ are in E, $1 \leq i \leq n$, and the V_i are distinct except for V_1 and V_{n+1}, which are equal.

Graph **G1** shown in Figure 4.11a contains the Hamiltonian cycle 1, 2, 8, 7, 6, 5, 4, 3, 1. Graph **G2** shown in Figure 4.11b contains no Hamiltonian cycle.

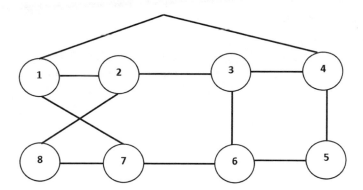

FIGURE 4.11A Graph G1

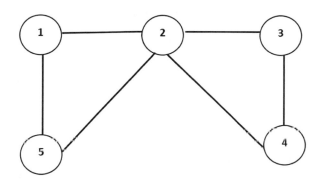

FIGURE 4.11B Graph G2

ALGORITHM FOR FINDING ALL HAMILTONIAN CYCLES

```
void Hamiltonian(int k)
// This program uses the recursive formulation of backtracking to find
//all the Hamiltonian cycles of a graph. The graph is stored as an adjacency
// matrix G[l:n] [l:n]. All cycles begin at node 1.
{
        do { // Generate values for x[k].
        NextValue(k); // Assign a legal next value to x[k].
        if (!x[k]) return;
        if (k == n) {
        for (int i=l; i<=n; i++) cout <<x[i] << ' ';
        cout <<"1\n";
        }
        else Hamiltonian(k+1);
        } while (1);
}
```

PROGRAM: GENERATING A NEXT VERTEX

```
void NextValue(int k)
//x[i], . . . , x[k-i] is a path of k-i distinct vertices. If x[k]==0, then no vertex
//has as yet been assigned to x[k]. After execution x[k] is assigned to the
// next highest numbered vertex which (1) does not already appear in
// x[1], x[2], . . . , x[k – 1]; and (2) is connected by an edge to x[k – 1]. Otherwise
    x[k]==0.
//If k==n, then in addition x[k] is connected to x[1].
{
        do {
        x[k] = (x[k]+i) % (n+i);        // Next vertex
        if (!x[k]) return;
        if (G[x[k-i]] [x[k]]) { // Is there an edge?
        for (int j=1; j<=k-i; j++) if (x[j]==x[k]) break;
                                        // Check for distinctness.
                if (j==k)        // If true, then the vertex is distinct.
                if ((k<n) ||((k==n) && G[x [n]] [x [1]]))
                return;
                }
        } while (1) ;
}
```

There is no known easy way to determine whether a given graph contains a Hamiltonian
cycle. A backtracking algorithm is used to find all the Hamiltonian cycles in a graph.
The graph may be directed or undirected. Only distinct cycles are output.

The backtracking solution vector $(x_1, ... , x_n)$ is defined so that **Xi** represents
the **i**th visited vertex of the proposed cycle. The function **NextValue (k)** is used to
determine a possible next vertex for the proposed cycle.

Example
Consider the graph shown in Figure 4.12.

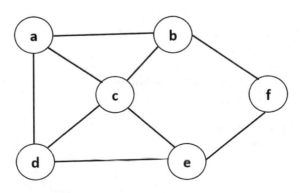

FIGURE 4.12   Example graph

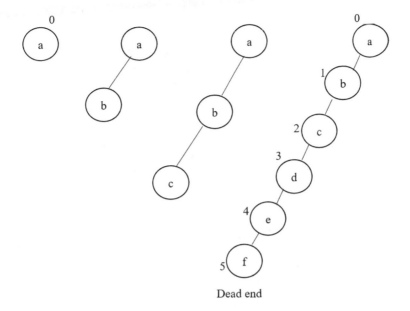

FIGURE 4.13 State-state tree

Start at vertex **a**; make vertex **a** the root of the state-space tree. Select vertex **b**. Next select vertex **c**. Then select **d**, then **e**, and finally **f**, which proves to be a dead end as shown in Figure 4.13.

So the algorithm backtracks from **f** to **e**, there are no useful alternatives. Backtrack from **e** to **d** and then to **c**, which provides the first alternative for the algorithm to pursue. Going from **c** to **e** eventually proves useless as shown in Figure 4.14.

The algorithm backtracks from **e** to **c**. There are no other alternatives. Then backtrack from **c** to **b** as shown in Figure 4.15. From there, it goes to the vertices **f**, **e**, **c**, and **d**, from which it returns to **a**, yielding the Hamiltonian circuit **a, b, f, e, c, d, a**.

Note

The traveling salesperson problem is used to find a tour that has minimum cost. This tour is a Hamiltonian cycle. For the simple case of a graph whose edge costs are all identical, Hamiltonian will find a minimum-cost tour if a tour exists. If the common edge cost is **c**, the cost of a tour is **cn** since there are **n** edges in a Hamiltonian cycle.

4.6 0/1 KNAPSACK (USING BACKTRACKING)

Given **n** positive weights W_i, **n** positive profits P_i, and a positive number 'm' that is the knapsack capacity, find a subset of the weights such that

$$\sum_{1 \leq i \leq n} W_i X_i \leq m \quad and \quad \sum_{1 \leq i \leq n} P_i X_i \text{ is maximized.}$$

The X_i's constitute a zero-one-valued vector. The solution space for this problem consists of the 2^n distinct ways to assign zero or one values to the X_i's.

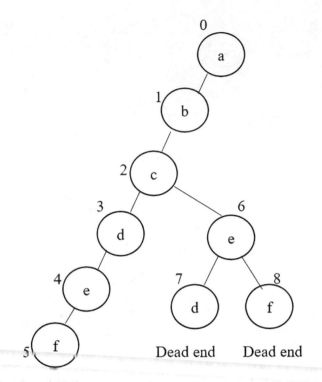

FIGURE 4.14   No Hamiltonian circuit

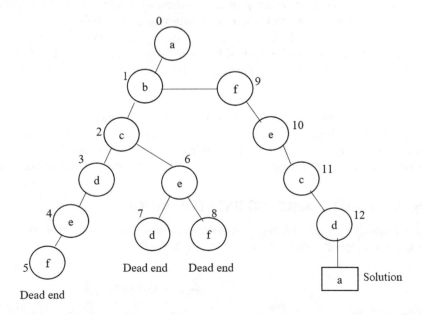

FIGURE 4.15   Hamiltonian circuit

A BOUNDING FUNCTION

```
float Bound(float cp, float cw, int k)
// cp is the current profit total, cw is the current
// weight total; k is the index of the last removed
// item; and m is the knapsack size.
{
        float b = cp, c = cw;
        for (int i=k+1; i<=n; i++) {
        c += w [i] ;
        if (c < m) b += p[i] ;
        else return(b + (1-(c-m)/w[i])*p[i]);
        }
        return(b);
}
```

From Bound, it follows that the bound for a feasible left child of a node Z is the same as that for Z. Hence, the bounding function need not be used whenever the backtracking algorithm makes a move to the left child of a node. The resulting algorithm is BKnap. It was obtained from the recursive backtracking schema.

So far, all our backtracking algorithms have worked on a static state-space tree. We now see how a dynamic state-space tree can be used for the knapsack problem. One method for dynamically partitioning the solution space is based on trying to obtain an optimal solution using the greedy algorithm.

We first replace the integer constraint $X_i = 0$ **or** **1** by the constraint $0 \le x_i \le 1$. This yields the relaxed problem

$$\max \sum_{1 \le i \le n} P_i X_i \ subject \ to \ \sum_{1 \le i \le n} W_i X_i \le m.$$

BACKTRACKING SOLUTION TO THE 0/1 KNAPSACK PROBLEM

```
void BKnap(int k, float cp, float cw)
// m is the size of the knapsack; n is the number of weights and profits. //w[]
    and p[] are the weights and profits. p[i]/w[i] >= p[i+l]/w[i+l].
// fw is the final weight of knapsack; fp is the final maximum profit. //x[k]==O
    if w[k] is not in the knapsack else x[k]==l.
{       // Generate left child.
        if (cw+w[k] <= m) {
        y[k]=l; if (k<n) BKnap(k+l, cp+p[k], cw+w[k]);
        if ((cp+p[k] > fp)&&(k==n)) {
        fp=cp+p[k]; fw=cw+w[k];
        for (int j=l;j<=k;j++) x[j]=y[j];
        }
        }
        //Generate right child.
        if (Bound(cp, cw, k) >= fp) {
```

```
y[k]=O; if (k<n) BKnap(k+l,cp,cw);
if ((cp > fp)&&(k == n)) {
fp = cp; fw = cw;
for (int j=l; j<=k; j++) x[j]=y[j];
}
}
}
```

Example

Consider the knapsack instance $n = 3$, $w = [20, 15, 15]$, $P = [40, 25, 25]$, and $C = 30$.

STATE-SPACE TREE

Best solution is L node. Profit is 50.

- The root is the only live node at this time. It is also the *E-node* (expansion node).
- From here, move to either B or C. Suppose the move is to B. The live nodes are A and B.
- B is the current *E*-node. At node B the remaining capacity **r** is 10, and the profit earned **cp** is 40. From B, move to either D or E. The move to D is infeasible, as the capacity needed to move there is $w_1 - 15$ The move to E is feasible, as no capacity is used in this move.
- *E* becomes the new *E-node*. The live nodes at this time are A, B and E. At node E, $r = 10$ and **cp = 40**. From E, move to either J or K. The move to node J is infeasible.

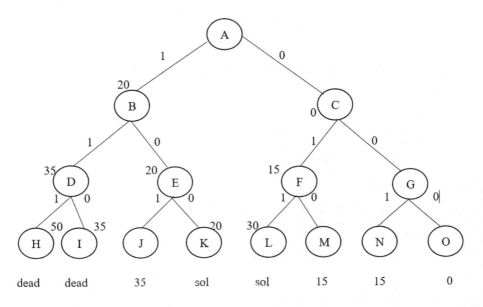

FIGURE 4.16   State-state tree

- Node K becomes the new *E-node*. Since K is a leaf, it is a feasible solution. This solution has a profit value $cp = 40$. The values of x are determined by the path from the root to k. The path is (A, B, E, K). Since node k cannot be expanded further, this node dies and back up to E node E also dies. Next back up to B, which also dies, and A becomes the *E-node* again.
- Now A is an E-node. It can be expanded further, and node C is reached. Now $r = 30$ and $cp = 0$. From C, move to either F or G. Suppose the move is to F; F becomes the new *E-node*, and the live nodes are A, C, and F.
- At F, $r = 15$ and $CP = 25$. From F, move to either L or M. Suppose the move is to L; now $r = 0$ and $cp = 50$. Since L is a leaf and it represents a better feasible solution than the best found so far (i.e., the one at node K), remember that this feasible solution is the best solution. Node L dies, back up to node F. Continuing in this way, search the entire tree. The best solution found during the search is the optimal one as shown in Figure 4.16.

5 Graph

5.1 INTRODUCTION

Graphs are formal descriptions for a wide range of familiar situations. The most common example is a road map, which shows the location of intersections and the roads that run between them. In graph terminology, the intersections are the nodes of the graph, and the roads are the edges. Sometimes our graphs are directed (like one-way streets) or weighted with a travel cost associated with each edge (like toll roads). As we give more details on the terminology and concepts in graphs, the similarity to road maps will be even clearer.

There are times when information needs to be distributed to a large number of people or to the computers on a large network. We would like this information to get everywhere, but we also don't want it going any place twice. Some groups of people will accomplish this by setting up a 'phone tree' where each person has a small number of people to call to pass on news. If everyone appears once in the tree and if the tree is not very deep, information will travel to everyone very quickly. For graphs, this is a bit more complicated, because there are typically many more connections between nodes than in a tree. We will look at two graph traversal methods, depth-first and breadth-first, to accomplish this.

Graph G consists of two sets V and E. The set V is a finite, nonempty set of vertices. The set E is a set of pairs of vertices; these pairs are called edges. A *graph* is a finite set of vertices (nodes) with edges between nodes, that is, Graph G = (V, E), where

$$V = \text{set of vertices and}$$
$$E = \text{set of edges.}$$

TYPES OF GRAPHS

There are two types of graphs:

1. Undirected
2. Directed

Undirected graph: In an undirected graph the pair of vertices representing any edge is unordered. A graph G is called **undirected** if every edge in it is undirected. Thus, the pairs (u, v) and (v, u) represent the same edge. The graph depicted in Figure 5.1 has six vertices and seven undirected edges:

$$V = \{a, b, c, d, e, f\}, E = \{(a, c), (a, d), (b, c), (b, f), (c, e), (d, e), (e, f)\}.$$

DOI: 10.1201/9781003355403-5

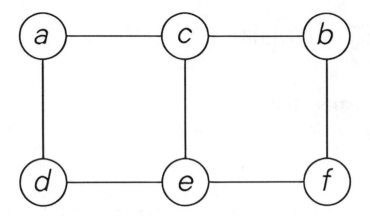

FIGURE 5.1   Undirected graph

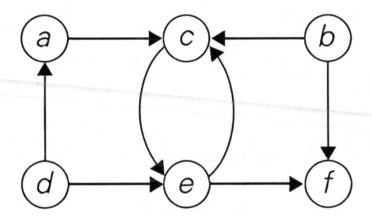

FIGURE 5.2   Directed graph

Directed graph: In a directed graph, each edge is represented by a directed pair
(u, v). Here **u** is the tail, and **v** is the head of the edge. Therefore (u, v) and (v, u) represent two different edges. A graph whose every edge is directed is called *directed*.
Directed graphs are also called *digraphs*. The digraph depicted in Figure 5.2 has six
vertices and eight directed edges:

$$V = \{a, b, c, d, e, f\},\ E = \{(a, c), (b, c), (b, f), (c, e), (d, a), (d, e), (e, c), (e, f)\}.$$

Some of the restrictions on graphs follow:

- A graph may not have an edge from a vertex v back to itself. That is, the
  edges of the form (v, v) are not legal. Such edges are known as self-edges
  or self-loops.
- A graph may not have multiple occurrences if the same edge.

Graph   **137**

Note: The number of distinct unordered pairs (u, v) with $u \neq v$ in a graph with n vertices is $\dfrac{n(n-1)}{2}$. This is the maximum number of edges in any n-vertex undirected graph. An n-vertex, undirected graph with exactly $\dfrac{n(n-1)}{2}$ edges is said to be complete.

TERMINOLOGY

Adjacency: Two vertices are adjacent if they are connected with an edge. When **(x, y)** is an edge, we say that **x** is adjacent to **y**, and **y** is adjacent from **x**.

Path: A path in Graph G is a sequence of vertices (nodes) $V_1, V_2, \ldots, V_K$ such that there is an edge from each node to the next one in the sequence $(V_i, V_{i+1}) \in E, 0 < i < n.$

Length of path: The length of a path is defined as the number of edges in a path.

Degree: The *degree* of a vertex is the number of edges incident to that vertex.

Let G be a directed graph.

- The *in degree* of a node x in G is the number of edges coming to x
- The *out degree* of x is the number of edges leaving x.

Let G be an undirected graph.

- The *degree* of a node x is the number of edges that have x as one of their end nodes.
- The *neighbors* of x are the nodes adjacent to x

Connected graph: An undirected graph is said to be *connected* iff for every pair of distinct vertices **u** and **v** in V(G), there is a path from v to v in G. Otherwise, the graph is *disconnected*.

Subgraph: A subgraph of G is a graph G' such that $V(G') \subseteq V(G)$ and $E(G') \subseteq E(G)$.

Connected component: A *connected component* of an undirected is a maximal connected subgraph.

Spanning tree: A *spanning tree* of an undirected Graph G is a subgraph of G that is a tree containing all the vertices of G.

GRAPH REPRESENTATION

There are two standard ways to represent a graph:

- Adjacency matrix
- Adjacency list

Adjacency Matrix Representation

Let G = (V, E) be a graph with n vertices, $n \geq 1$. In this representation, each graph of *n* nodes is represented by an $n \times n$ matrix. Consider a graph with connected components as shown in Figure 5.3.

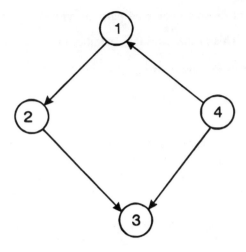

FIGURE 5.3   A graph with connected components

The adjacency matrix of G is a two-dimensional array with the property that

$$A[i\ j] = 1 \text{ iff } (i, j) \text{ is an edge.}$$
$$A[i, j] = 0 \text{ iff } (i, j) \text{ is not an edge}$$

The adjacency matrix is $\begin{bmatrix} 0 & 1 & 0 & 0 \\ 0 & 0 & 1 & 0 \\ 0 & 0 & 0 & 0 \\ 1 & 0 & 1 & 0 \end{bmatrix}$.

Advantages
- Simple to implement
- Easy and fast to tell if a pair (i, j) is an edge: simply check if $A[i][j]$ is 1 or 0.

Disadvantage
- No matter how few edges the graph has, the matrix takes $O(n^2)$ in memory.

Adjacency List Representation
In this representation of graphs, the n rows of the adjacency matrix are represented as n-linked lists. There is one list for each vertex in G. The nodes in the list I represent the vertices that are adjacent from vertex i.

Each node has two fields: *vertex* and *link*. The vertex field contains the indices of the vertices adjacent to vertex i. Each list has a head node. The head nodes are sequential and so provide easy random access to the adjacency list for any particular vertex. An adjacency list representation is shown in Figure 5.4.

Graph 139

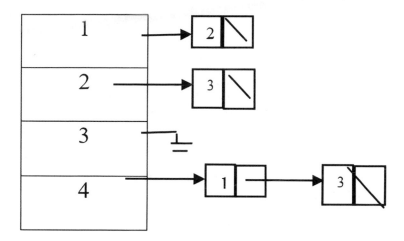

FIGURE 5.4   Adjacency list representation

Advantage: It saves memory space.
Disadvantage: It can take up to O(n) time to determine if a pair of nodes (i, j) is an edge: one would have to search the linked list L[i], which takes time proportional to the length of L[i].

5.2   GRAPH TRAVERSALS

Traversing a graph is a process of visiting each node in a graph exactly once and in a systematic manner. Different nodes of a graph may be visited, possibly more than once, during a traversal or search. If the search results in a visit to all the vertices, it is called traversal.

There are two standard graph search and traversal techniques:

- Breadth-first search (BFS) and breadth-first traversal (BFT)
- Depth-first search (DFS) and depth-first traversal (DFT)

BFS AND BFT

- In a BFS, we start at a vertex *v* and mark it as having been reached (visited). The vertex *v* is at this time said to be unexplored.
- A vertex is said to have been explored by an algorithm when the algorithm has visited all vertices adjacent from it. All unvisited vertices adjacent from *v* are visited next. These are new unexplored vertices.
- Vertex *v* has now been explored. The newly visited vertices haven't been explored and are put onto the end of a list of unexplored vertices.
- The first vertex on this list is the next to be explored. Exploration continues until no unexplored vertex is left.

- The list of unexplored vertices operates as a queue and can be represented using any of the standard queue representations.

Program: BFS

```
Void BFS (int v)
// A breadth-first search of G is carried out beginning at vertex v. //For any
    node i, visited [i]==1 if 'i' has already been visited. The //graph G and
    array visited [] are global; visited [] is initialized to zero.
{
    int u=v; Queue q[SIZE];
            // q is a queue of unexplored vertices.
    visited[v] = 1;
    do {
            for all vertices w adjacent from u {
            if (visited[w] == 0) {
            q.AddQ(w);        // w is unexplored,
            visited[w]=1;
            }
        }
        if (q.Qempty()) return;    // No unexplored vertex
        q.DeleteQ(u);              // Get first unexplored vertex.
    } while (1);
}
```

If a BFS is used on a connected undirected graph G, then all vertices in G get visited, and the graph is traversed. However, if G is not connected, then at least one vertex of G is not visited. A complete traversal of the graph can be made by repeatedly calling BFS each time with a new unvisited starting vertex. The resulting traversal algorithm is known as a BFT. Consider the example graph shown in Figure 5.5 and how to find out the adjacency matrix and BFS in Figure 5.6.

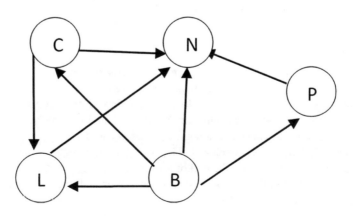

FIGURE 5.5   An example graph

Graph 141

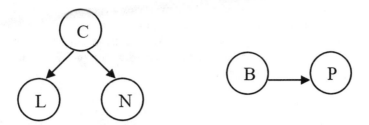

FIGURE 5.6 BFS

Program: Breadth-First Graph Traversal

```
void BFT(struct treenode G[], int n)
// Breadth first traversal of G
{
        int i; boolean visited[SIZE];
        for (i=l; i<=n; i++) // Mark all vertices unvisited.
        visited[i] = 0;
        for (i=l; i<=n; i++)
        if (!visited[i]) BFS(i);
}
```

Adjacency Matrix

	C	L	N	B	P
C	0	1	1	0	0
L	0	0	1	0	0
N	0	0	0	0	0
B	1	1	1	0	1
P	0	0	1	0	0

Procedure

1. Begin with any node and mark it as visited.
2. Proceed to the next node having any edge connection to the node in step 1 and mark it as visited.
3. Come back to the node in step 1; descend along an edge toward an unvisited node and mark the new node as visited.
4. Repeat step 3 until all the nodes adjacent to the node in step 1 have been marked as visited.
5. Repeat steps 1 through 4 from the node visited in step 2; then start from the node visited in step 3.
6. Keep this up as long as possible.

DFS AND DFT

- A DFS of a graph differs from a BFS in that the exploration of a vertex v is suspended as soon as a new vertex is reached.

- At this time, the exploration of the new vertex *u* begins. When this new vertex has been explored, the exploration of *v* continues.
- The search terminates when all reached vertices have been fully explored.
- A DFT of a graph is carried out by repeatedly calling DFS, with a new unvisited starting vertex each time.

Difference Between BFS and DFS

- In a BFS, a node is fully explored before the exploration of any other node begins. The next node to explore is the first unexplored node remaining. The exercises examine a search technique (D-search) that differs from BFS only in that the next node to explore is the most recently reached unexplored node.
- In a DFS, the exploration of a node is suspended as soon as a new unexplored node is reached. The exploration of this new node is immediately begun. An example is shown in Figure 5.7, and a DFS is shown in Figure 5.8.

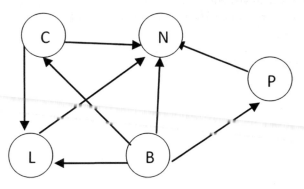

FIGURE 5.7   An example graph

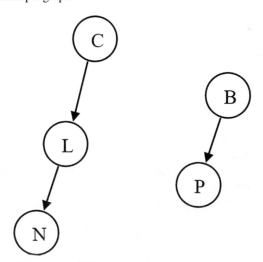

FIGURE 5.8   DFS

Graph   143

Program: DFS of a Graph

```
void DFS(int v)
// Given an undirected (directed) graph G = (V, E) with
// n vertices and an array visited[] initially set
// to zero, this algorithm visits all vertices
// reachable from v. G and visited[] are global.
{
visited[v] = 1;
for each vertex w adjacent from v {
if (!visited[w]) DFS(w);
}
```

Adjacency Matrix

	C	L	N	B	P
C	0	1	1	0	0
L	0	0	1	0	0
N	0	0	0	0	0
B	1	1	1	0	1
P	0	0	1	0	0

Procedure

1. Choose any node in the graph designated as a search node and mark it as visited.
2. Using the adjacency matrix of the graph, find a node adjacent to the search node that has not been visited, designate it as a new search node, and mark it as visited.
3. Repeat the step 2 using the new search node. If no node satisfying the step 2 can be found, then return to the previous search node and continue from there.
4. When a return to a previous search node in step 3 is impossible, then the search from the originally chosen node is complete.
5. If the graph still contains unvisited nodes, then choose any node that has not been visited and repeat steps 1 through 4.

5.3   CONNECTED COMPONENTS AND SPANNING TREES

A graph is said to be connected if for every pair of its vertices **u** and **v** there is a path from **u** to **v**. If a graph is not connected, such a model will consist of several connected pieces that are called connected components of the graph; that is, a connected component is the maximal subgraph of a given graph.

EDGE CLASSIFICATION OF DFS

DFS introduces an important distinction among the edges in the graph. They are four edge types produced by a DFS on G in terms of a depth-first forest.

1. Tree edge

   Edge (*u*, v) is a *tree edge* if vertex v is first discovered by exploring edge (*u*, *v*) as shown in Figure 5.9.

2. Back edge

   Edge (u, v) is a *back edge* that connects a vertex *u* to ancestor v in a depth-first tree as shown in Figure 5.10.

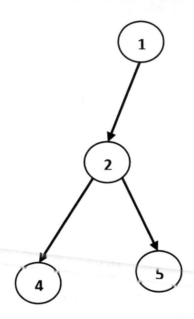

FIGURE 5.9   Tree edge

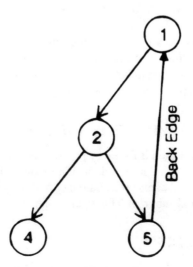

FIGURE 5.10   Back edge

Graph   145

3. Forward edge

Edge (u, v) is a forward edge (or non-tree edge) that connects a vertex *u* to descendent **v** in a depth-first tree as shown in Figure 5.11.

4. Cross edge

Cross edges are other edges that go between
- vertices of the same tree, as long as one vertex is not an ancestor of the other.
- vertices of different trees as shown in Figure 5.12.

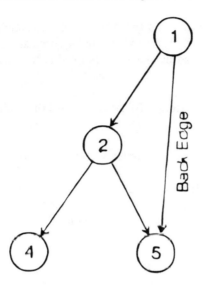

FIGURE 5.11   Forward edge

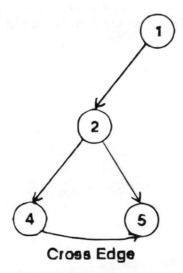

Cross Edge

FIGURE 5.12   Cross edge

Applications

1. A BFS is used to determine whether G is a connected graph or not because
   - if G is connected, then all vertices will be visited on the first call to BFS.
   - if G is not connected, then at least two calls to BFS will be needed.
2. A BFT is used to obtain the connected component of a graph.

All newly visited vertices on a call to BFS from BFT represent the vertices in a connected component of G.

- Connected components of a graph can be found using a BFT.
- Modify the BFS so that all newly visited vertices are put onto a list.
- The subgraph formed by the vertices on this list makes up a connected component.

If adjacency lists are used, a BFT will obtain the connected components in $\theta(n + e)$ time.

3. A BFS is used to obtain the reflexive transitive closure matrix of an undirected graph.
4. A BFS is used to determine the existence of a spanning tree.
   - A graph G is a spanning tree iff G is connected.
   - The computed spanning tree is not a minimum spanning tree.
5. A DFT is used to obtain the connected components as well as the spanning tree.

5.4 SPANNING TREES

The spanning trees using a BFS (or DFS) can be easily obtained. The spanning trees given by a BFS and a DFS are not identical. Figure 5.13 gives an example of this algorithm in operation.

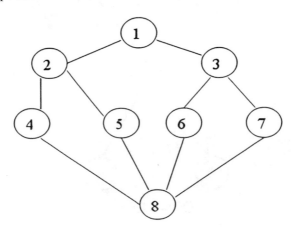

FIGURE 5.13   An example graph

Graph 147

BREADTH-FIRST SPANNING TREE

The spanning trees obtained using a BFS are called *breadth-first spanning trees* as shown in Figure 5.14.

DEPTH-FIRST SPANNING TREE

The spanning trees obtained using a DFS are called *depth-first spanning trees* as shown in Figure 5.15.

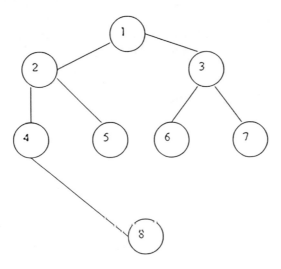

FIGURE 5.14   Breadth-first spanning tree

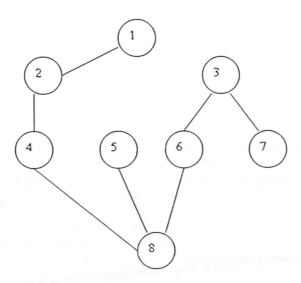

FIGURE 5.15   Depth-first spanning tree

MINIMUM SPANNING TREE

A *spanning tree* of an undirected connected graph is its connected acyclic subgraph (i.e., a tree) that contains all the vertices of the graph. If such a graph has weights assigned to its edges, a *minimum spanning tree* (MST) is its spanning tree of the smallest weight, where the *weight* of a tree is defined as the sum of the weights on all its edges. The *MST problem* is the problem of finding an MST for a given weighted connected graph. A spanning tree **S** is a subset of a tree **T** in which all the vertices of tree **T** are present, but it may not contain all the edges.

Let G = (V, E) be an undirected connected graph. A subgraph $t = (V, E')$ of G is a *spanning tree* of G iff t is a tree.

The spanning tree of a weighted connected graph **G** is called an MST if its weight is minimum. The number of edges in the MST is |V| − 1. The MST is a tree because it is acyclic; it is spanning because it covers with minimum cost.

MST problems can be solved using the greedy technique. Two algorithms are used to construct an MST for an undirected graph:

1. Prim's algorithm
2. Kruskal's algorithm

PRIM'S ALGORITHM

Prim's algorithm constructs a minimum spanning tree through a sequence of expanding subtrees. The initial subtree in such a sequence consists of a single vertex selected arbitrarily from the set V of the graph's vertices. On each iteration, the algorithm expands the current tree in a greedy manner by simply attaching to it the nearest vertex that is not in that tree. (By the nearest vertex, we mean a vertex that is not in the tree connected to a vertex in the tree by an edge of the smallest weight. Ties can be broken arbitrarily.) The algorithm stops after all the graph's vertices have been included in the tree being constructed. Since the algorithm expands a tree by exactly one vertex on each of its iterations, the total number of such iterations is $n - 1$, where n is the number of vertices in the graph. The tree generated by the algorithm is obtained as the set of edges used for the tree expansions. One way to compute an MST is to grow the tree in successive stages. In each stage, one node is picked as the root and an edge is added; that is, an associated vertex is added to the tree until all the vertices are present in the tree with |V − 1| edges. Figure 5.16 gives an example of this algorithm in operation.

A greedy method to obtain a minimum-cost spanning tree builds this tree edge by edge. The next edge to include is chosen according to some optimization criterion.

The Strategy

1. One node is picked as a root node (**u**) from the given connected graph.
2. At each stage choose a new vertex **v** from **u**, by considering an edge (**u, v**) with minimum cost among all the edges from **u**, where **u** is already in the tree and **v** is not in the tree.
3. The Prim's algorithm table is constructed with three parameters:
   • Known—the vertex is added in the tree or not.

Graph   149

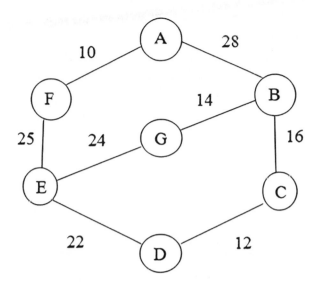

FIGURE 5.16 The original graph

- dv—the weight of the shortest arc connecting **v** to a known vertex
- pv—last vertex that causes a change in dv.
4. After selecting the vertex **v**, then update rule is applied for each unknown **w** adjacent to **v**. The rule is **dw = min(dw, C'$_{w,v}$)**; that is, if more than one path exists between **v** to **w**, then **dw** is updated with minimum cost.

The initial configuration of a table for the given example follows:

v	known	dv	Pv
A	0	0	0
B	0	∞	0
C	0	∞	0
D	0	∞	0
E	0	∞	0
F	0	∞	0
G	0	∞	0

The table after A is declared as 'known'

v	known	dv	pv
A	1	0	0
B	0	28	A
C	0	∞	0
D	0	∞	0
E	0	∞	0
F	0	10	A
G	0	∞	0

Known(A) = 1, dv(B) = 28 cost from A to B, pv(B) = A the vertex to reach B.
  The table after F is declared as 'known'

v	known	dv	pv
A	1	0	0
B	0	28	A
C	0	∞	0
D	0	∞	0
E	0	25	F
F	1	10	A
G	0	∞	0

The table after E is declared as 'known'

v	known	dv	pv
A	1	0	0
B	0	28	A
C	0	∞	0
D	0	22	E
E	1	25	F
F	1	10	A
G	0	24	E

The table after D is declared as 'known'

v	known	dv	pv
A	1	0	0
B	0	28	A
C	0	12	D
D	1	22	E
E	1	25	F
F	1	10	A
G	0	24	E

The table after C is declared as 'known'

v	known	dv	pv
A	1	0	0
B	0	16	C
C	1	12	D
D	1	22	E
E	1	25	F
F	1	10	A
G	0	24	E

Here the **dv** value of B is updated using an update rule. That is, B can be reached through either A or C from the edge with the minimum cost that is chosen.

Graph 151

The table after B is declared as 'known'

v	known	dv	pv
A	1	0	0
B	1	16	C
C	1	12	D
D	1	22	E
E	1	25	F
F	1	10	A
G	0	14	B

In this table, the **dv** value of G is updated using an update rule.
The table after G is declared as 'known'

v	known	dv	pv
A	1	0	0
B	1	16	C
C	1	12	D
D	1	22	E
E	1	25	F
F	1	10	A
G	1	14	B

The edges in the spanning tree can be read from the table:

(A, F)(F, E)(E, D)(D, C)(C, B)(B, G).

The total number of edges is 6, which is equivalent to |V| − 1 (number of vertices − 1).
The sequence of the edges added at each stage starts from vertex A and ends at vertex
G. Figure 5.17 and Figure 5.18 shows the application of Prim's algorithm.

Prim's Minimum-Cost Spanning Tree Algorithm

float Prim(int E[] [SIZE], float cost[] [SIZE], int n, int t[] [2])
*//E is the set of edges in G. cost[l:n] [l:n] is the cost adjacency matrix of an n
    vertex //graph such that cost[i][j] is either a positive real no. or infinity if
    no edge (i, j) exists*
// A minimum spanning tree is computed and stored as a set of edges in the
*// array t[1:n − 1] [1:2].(t[i][1], t[i][2]) is an edge in the minimum-cost span-
    ning tree.*
// The final cost is returned.
{
        int near [SIZE], j, k, l;
        let (k,l) be an edge of minimum cost in E;
        float mincost = cost[k] [l];
        t[1] [1] = k; t[l][2] = 1;
        for (int i=l; i<=n; i++) // *Initialize near.*

```
if (cost[i][l] < cost [i][k]) near [i] = 1;
else near[i] = k;
near[k] = near[l] = 0;
for (i=2; i <= n-l; i++) { // Find n – 2 additional
                          // edges for t.
let j be an index such that near[j] !=0 and
cost[j] [near[j]] is minimum;
t [i][l] = j; t [i][2] = near [j] ;
mincost = mincost + cost[j] [near[j]];
near[j]=0;
for (k=l; k<=n; k++) // Update near[].
if ((near[kJ!=O) &&
(cost[k][near[k]]>cost [k][j]))
near[k]= j;
}
return(mincost);
}
```

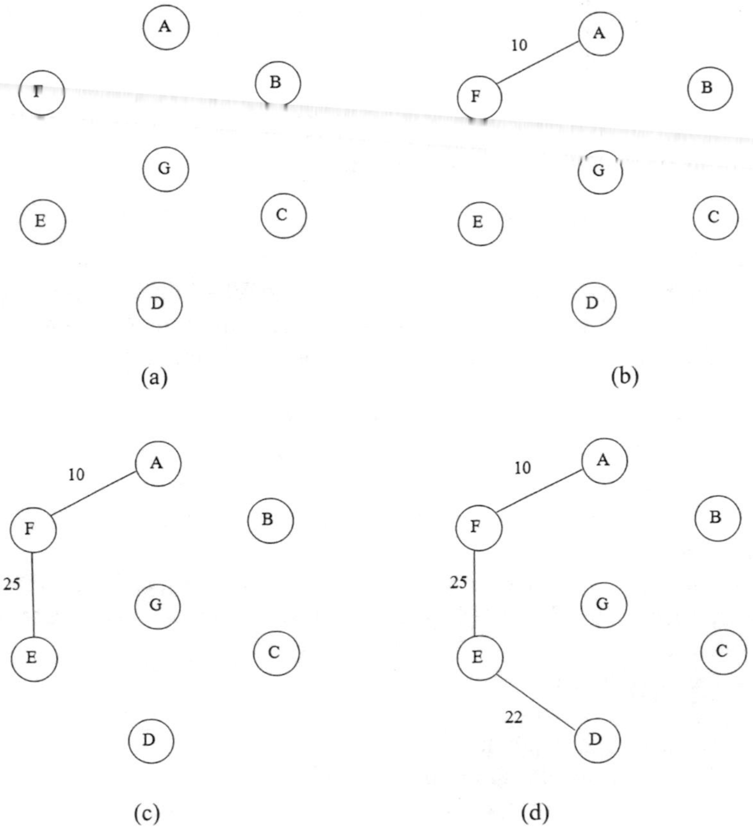

FIGURE 5.17   Stages in Prim's algorithm

Graph   153

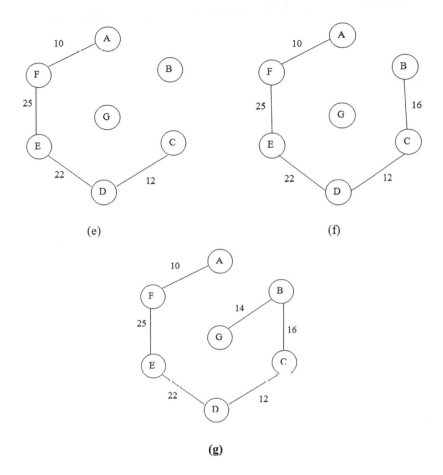

(e)

(f)

(g)

FIGURE 5.18   Stages in Prim's algorithm

KRUSKAL'S ALGORITHM

This is another greedy strategy for selecting edges in order of the smallest weight and accepting an edge if it does not cause a cycle. Kruskal's algorithm maintains a forest collection of trees. It is named ***Kruskal's algorithm***after Joseph Kruskal, who discovered this algorithm when he was a second-year graduate student. Kruskal's algorithm looks at an MST of a weighted connected graph $G = |V, E|$ as an acyclic subgraph with $|V| - 1$ edges for which the sum of the edge weights is the smallest.

(It is not difficult to prove that such a subgraph must be a tree.) Consequently, the algorithm constructs an MST as an expanding sequence of subgraphs that are always acyclic but are not necessarily connected in the intermediate stages of the algorithm.

The algorithm begins by sorting the graph's edges in a nondecreasing order of their weights. Then, starting with the empty subgraph, it scans this sorted list, adding the next edge on the list to the current subgraph if such an inclusion does not create a cycle and simply skipping the edge otherwise.

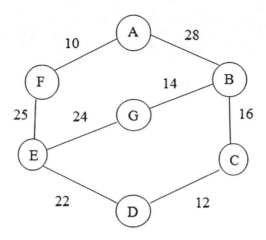

FIGURE 5.19 The original graph

Initially there are |V| single-node trees. Adding an edge merges two trees into one. When the algorithm terminates there is only one tree, and this is the MST. The algorithm terminates when enough edges are accepted. It turns out to be simple to decide whether edge (u, v) should be accepted or rejected. Two vertices belong to the same set if and only if there are connected in the current spanning forest. Each vertex is initially in its own set. If **u** and **v** are in the same set, the edge is rejected because they are already connected or because adding (u, v) would form a cycle. Otherwise, the edge is accepted, and a union is performed on the two sets containing **u** and **v**. Figure 5.19 gives an example of this algorithm in operation, and the stages in Kruskal's algorithm are shown in Figure 5.20.

THE STRATEGY

1. The edges are built into a min-heap structure, and each vertex is considered a single-node tree.
2. The Deletemin operation is utilized to find the minimum-cost edge (u, v).
3. The vertices u and v are searched in the spanning tree set S, and if the returned sets are not the same, then (u, v) is added to the set S with the constraint that adding (u, v) will not create a cycle in the spanning tree set S.
4. Repeat steps 2 and 3 until a spanning tree is constructed with |V| − 1 edges.

Edge	Weight	Action
(A, F)	10	Accepted
(C, D)	12	Accepted
(B, G)	14	Accepted
(B, C)	16	Accepted
(D, E)	22	Accepted
(E, G)	24	Rejected because it forms a cycle
(F, E)	25	Accepted
(A, B)	28	Rejected

Graph   155

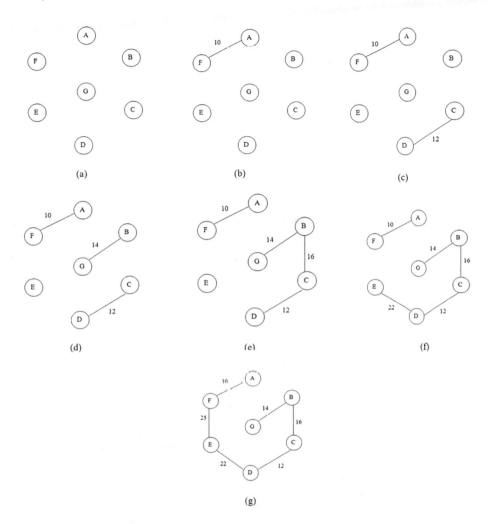

FIGURE 5.20 Stages in Kruskal's algorithm

float Kruskal(int E[][SIZE], float cost[] [SIZE], int n, int t [][2])
// E is the set of edges in G. G has n vertices. cost[u] [v] is the cost of edge (u, v).
//t is the set of edges in the minimum-cost spanning tree. The final cost is
    returned.
{ int parent [SIZE];
        construct a heap out of the edge costs using Heapify;
        for (int i=1; i<=n; i++) parent[i] = −1;// Each vertex is in a different set.
        i = 0; float mincost = 0.0;
        while ((i < n−1) && (heap not empty)) {
        delete a minimum cost edge (u,v) from the heap and reheapify using
            Adjust;
        int j = Find(u); int k = Find(v);

```
if (j ! = k) { i++;
t[i] [1] = u; t[i] [2]=v;
mincost += cost[u] [v];
Union(j, k);
}
}
if (i ! = n–1) cout " "No spanning tree" " endl;
else return(mincost);
}
```

ANALYSIS

The worst-case running time of this algorithm is $O(|E|log|E|)$ with heap operations.

5.5   BICONNECTED COMPONENTS AND DFS

ARTICULATION POINT

A vertex **v** in a connected graph G is an **articulation point** if the deletion of vertex **v** together with all edges incident to **v** disconnects the graph into two or more non-empty components. Consider the connected graph in Figure 5.21.

In Figure 5.21, vertex 2 is an articulation point as the deletion of vertex 2 and edges (1, 2), (2, 3), (2, 5), (2, 7), and (2, 8) leaves behind the two disconnected non-empty components.

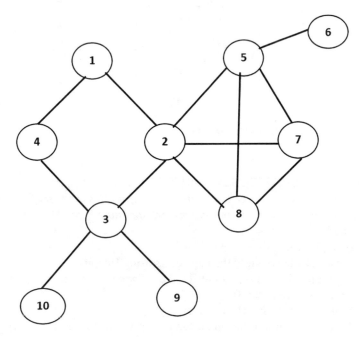

FIGURE 5.21   Connected graph

Graph   157

RESULT OF DELETING VERTEX 2

The graph shown in Figure 5.22 contains two other articulation points: vertices 5 and 3.

BICONNECTED GRAPH

A graph G is biconnected iff it contains no articulation points. The graph in Figure 5.23 is biconnected.

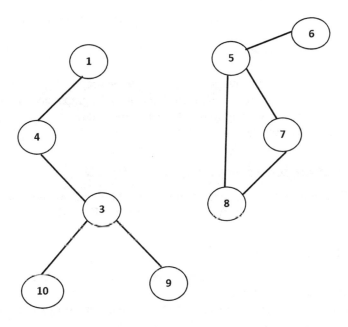

FIGURE 5.22   Connected graph with articulation points

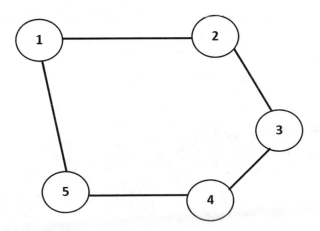

FIGURE 5.23   Biconnected graph

- The presence of an articulation point may be an undesirable feature in many cases.
- Consider a communications network with nodes as communication stations and edges as communication lines.
- A failure of a communication station may result in loss of communication between other stations as well if the graph is not biconnected.

ALGORITHM TO DETERMINE THE BICONNECTED COMPONENTS

The algorithm to determine if a connected graph is biconnected if it contains no articulation points.

If a graph is not biconnected,

- identify all the articulation points in a connected graph.
- determine a set of edges whose inclusion makes the graph biconnected.
  1. Find the maximal subgraphs of G that are biconnected.
  2. Find the biconnected component.

A MAXIMAL BICONNECTED SUBGRAPH

$G^1 = (V^1, E^1)$ is a maximal biconnected subgraph of G if and only if G has no biconnected subgraph $G^{11} = (V^{11}, E^{11})$ such that $V^1 \subset V^{11}$ and $E^1 \subset E^{11}$.

BICONNECTED COMPONENT

A maximal biconnected subgraph is a biconnected component. Biconnected components are shown in Figure 5.24.

Note: Two biconnected components can have at most one vertex in common, and this vertex is an articulation point.

IDENTIFYING THE ARTICULATION POINTS AND BICONNECTED COMPONENTS

This section illustrates the use of the biconnected components algorithm on the simple undirected graph that is shown in Figure 5.25.

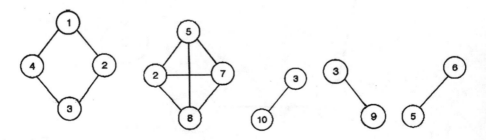

FIGURE 5.24   Biconnected subgraph

Graph   159

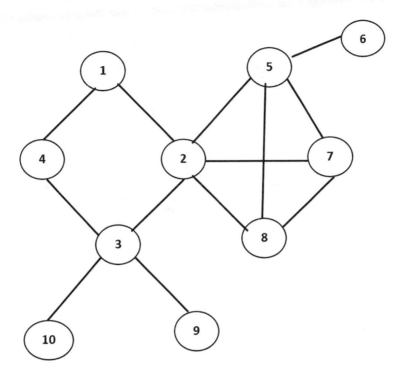

FIGURE 5.25   A simple undirected graph.

- Two biconnected components of the same graph can have at most one vertex in common.
- No edge can be in two or more biconnected components.
- The biconnected components of G partition the edges of G.
- The biconnected components of a connected, undirected graph G can be found by using any depth-first spanning tree of G.

Steps:

- A depth-first spanning tree has a property that is very useful in identifying articulation points and biconnected components. So find the depth-first spanning tree for the given graph G.
- The depth-first numbers are the numbers that give the order in which the DFS visits the vertices. The biconnected components are shown graphically in Figure 5.26.

dfn[1] = 1, dfn[4] = 2, dfn[3] = 3, dfn[4] = 4, dfn[1] = 5, dfn[4] = 6, dfn[1] = 7, dfn[4] = 8, dfn[1] = 9, dfn[4] = 10

Tree edges: The edges that form the depth-first spanning tree are called tree edges. A solid line in Figure 5.26 represents a tree edge.

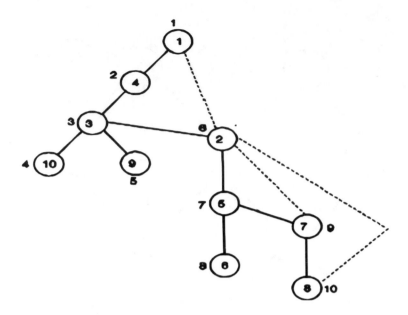

FIGURE 5.26   Biconnected components

Back edges: The remaining edges other than the tree edge in the graph are called back edges. A broken line in Figure 5.26 represents a back edge.

The root node of a depth-first spanning tree is an articulation point iff it has at least two children. Furthermore, if **u** is any other vertex, then it is not an articulation point iff from every child **w** of **u** it is possible to reach an ancestor of **u** using only a path made up of descendants of **w** and a back edge.

- For each vertex **u**, define L[u] is the lowest *dfn* that we can reach from **u** using a path of descendants followed by at most one back edge.

  L[u] = min {dfn[u], min{L[w] /w is a child u}, min {dfn[w] /(u, w) is a back edge}

- If **u** is not the root, then **u** is an articulation point iff **u** has a child **w** such that L[w] ≥ dfn[u].

Consider the depth-first spanning tree.
    L[u] can be easily computed if the vertices of the depth-first spanning tree are visited in post order.

    L [10] = min {4, -, -} = 4
    L [9] = min {5, -, -} = 5
    L [6] = min {8, -, -} = 8

Graph   161

L [8] = min {10, -, 6} = 6
L [7] = min {9, 6, 6} = 6
L [5] = min {7, min {8, 6}, -} = min {7, 6} = 6
L [2] = min {6, 6, min{9,10,1}} = min{6,6,1} = 1
L [3] = min {3, min{4,5,1}, -} = min{3,1} = 1
L [4] = min {2, 1, -} = 1
L [1] = min {1, 1, 6} = 1

Now find the articulation point by,

u = 1, L[w] ≥ dfn[u]
        L[4] ≥ dfn[1], 1 ≥ 1 but u = 1 is the root.
u = 2, L[w] ≥ dfn[u]
        L[5] ≥ dfn[2], 6 ≥ 6. Therefore, 2 is an articulation point.
u = 3, L[w] ≥ dfn[u]
        L[10] ≥ dfn[3], 4 ≥ 3. Therefore, 3 is an articulation point.
u = 4, L[w] ≥ dfn[u]
        L[3] ≥ dfn[4], 1! = 2.
u = 5, L[w] ≥ dfn[u]
        L[6] ≥ dfn[5], 8 ≥ 7. Therefore, 5 is an articulation point.

Hence, the articulation points are 2, 3, and 5.

Pscudo code to compute *dfn* and *L:*

```
void Art(int u, int v)
// u is a start vertex for DFS. v is its parent if any in the depth-first spanning
// tree. It is assumed that the global array dfn is initialized to zero and that
    the global
// variable num is initialized to 1. n is the number of vertices in G.
{
        extern int dfn[], L[], num, n;
        dfn[u] = num; L[u] = num; num++;
        for each vertex w adjacent from u {
        if (!dfn[w]) {
        Art(w, u); // w is unvisited
        L[u] = min(L[u], L[w]);
        }
        else if (w != v) L[u] = min(L[u], dfn[w]);
        }
}
```

PSEUDO CODE TO DETERMINE BICOMPONENTS

```
void BiComp(int u, int v)
// u is a start vertex for DFS. v is its parent if any in the depth-first spanning
// tree. It is assumed that the global array dfn is initialized to zero and that
    the global
```

```
// variable num is initialized to 1. n is the number of vertices in G. s is a stack
   of edges.
{
        extern int dfn[], L[], num, n;
        dfn[u] = num; L[u] = num; num++;
        for each vertex w adjacent from u {
        if ((v != w) && (dfn[w] < dfn[u]))
        add (u, w) to top of s;
        if (!dfn[w]) {
        if (L[w] >= dfn[u]) {
        cout << "New bicomponent\n";
        do {
        delete an edge from the top of stack s;
        let this edge be (x, y);
        cout<<"("<< x << "," <<y << ") ";
        } while (((x, y)! = (u, w))&& ((x, y) ! = (w, u)));
        }
        BiComp(w, u); // w is unvisited
        L[u] = min(L[u], L[w]);
        }
        else if (w != v) L[u] = min(L[u], dfn[w]);
        }
}
```

5.6   BRANCH AND BOUND

INTRODUCTION

Branch and bound refers to all state-space search methods in which all children of an
E-node are generated before any other live node can become the E-node. Depending
on the two graph-search strategies, branch-and-bound strategies are generalized into
two types:

1. BFS-like state-space search (or) FIFO (first-in, first-out) search
2. D-search-like state-space search (or) LIFO (last-in, first-out) search

BFS-like state-space search: A *BFS-like state-space search* is an FIFO search in
terms of live nodes. The FIFO search is the method that searches the tree using a
breadth-first search.

- List of live nodes is a queue.

D-search-like state-space search: A *D-search-like state-space search* is an LIFO
search in terms of live nodes. The LIFO search is the method that searches the tree
as in a BFS but replaces the FIFO queue with a stack.

- List of live nodes is a stack.

Graph   **163**

In both BFS and D-search, the exploration of a new node cannot begin until the node currently being explored is fully explored. Just like backtracking, the bounding functions to avoid generating subtrees that do not contain an answer node are used.

Like the backtracking, branch and bound uses the same terminology, which follows:

> **Live node:** A live node is a node that has been generated but whose children have not yet been generated.
>
> **E-node:** An E-node is a live node whose children are currently being explored. In other words, an E-node is a node currently being expanded.
>
> **Dead node:** A dead node is a generated node that is not to be expanded or explored any further. All children of a dead node have already been expanded.

EXAMPLE: 4-QUEENS

FIFO branch-and-bound algorithm

- Initially, there is only one live node; no queen has been placed on the chessboard.
- The only live node becomes the E node.
- Expand and generate all its children, children being a queen in columns 1, 2, 3, and 4 of row 1 (only live nodes left).
- The next E-node is the node with the queen in row 1 and column 1.
- Expand this node and add the possible nodes to the queue of live nodes.

Compared with backtracking algorithm, backtracking is a superior method for this search problem.

LEAST-COST SEARCH

In both LIFO and FIFO branch and bound, the selection rule for the next E-node is rigid. The Selection rule does not give preference to the nodes that will lead to an answer quickly, just to the queues those behind the current live nodes.

1. In the 4-queens problem, if three queens have been placed on the board, it is obvious that the answer may be reached in one more move.
2. The rigid selection rule requires that other live nodes be expanded, and then, the current node be tested.

Ranking Function

The search for an answer node can be speeded up by using a ranking function $c(.)$ for live nodes. The next E-node is selected on the basis of this ranking function.

Let $c(x)$ is the estimated additional computational effort or the additional cost needed to reach an answer node from the live node X. For any node x, the cost could be given by the number of nodes in subtree **x** that need to be generated before an answer can be reached. The search will always generate the minimum number of nodes.

1. Number of levels to the nearest answer node in the subtree x
   - **c(root)** for a 4-queens problem is 4.
   - The only nodes to become E-nodes are the nodes on the path from the root to the nearest answer node.

Problem

The problem with the previously discussed techniques for computing the cost at node **x** is that they involve the search of the subtree at **x**, implying the exploration of the subtree.

- By the time the cost of a node is determined, that subtree has been searched, and there is no need to explore **x** again.
- The preceding point can be avoided by using an estimate function **g(x)** instead of actually expanding the nodes.

Estimate Function

- Let g(x) be an estimate of the additional effort needed to reach an answer from node x.
- h(x) is the cost of reaching x from the root.
- f(.) is any nondecreasing function.

Node **x** is assigned a rank using a function **c(.)** such that

$$c(x) = f(h(x)) + g(x).$$

Least-Cost Branch-and-Bound Search

A search strategy that uses a cost function **c(x) = f(h(x)) + g(x)** to select the next E-node would always choose for its next E-node, a live node with least **c(.)**. Hence, such a strategy is called least-cost search or LC-search.

BFS and DFS are special cases of LC-search.

An LC-search, with g(x) = 0 and f(h(x)) = level of x is a BFS.
An LC-search with f(h(x) = 0 and g(x) ≥ g(y) whenever **y** is a child of **x** is a DFS. The LC-search with bounding functions is an LC branch-and-bound search.

Cost Function

The actual cost of x is denoted by c(x).

- If x is an answer node, c(x) is the cost of reaching x from the root of the state-space tree.

Graph   165

- If x is not an answer node, c(x) =∞, provided the subtree **x** contains no answer node.
- If subtree x contains an answer node, c(x) is the cost of a minimum-cost answer node in subtree x. **c(.)** with f(h(x)) = h(x) is an approximation to c(.), where h(x) is the cost of reaching x from the root.

BOUNDING

A branch-and-bound method searches a state-space tree using any search mechanism in which all children of the E-node are generated before another node becomes the E-node.

Each answer node x has a cost c(x), and we have to find a minimum-cost answer node.

Search Strategies

The three common search strategies include

1. LC,
2. FIFO, and
3. LIFO.

Cost Function

A cost function c(.) such that (x) ≤ c(x) is used to provide the lower bound on the solution obtainable from any node x.

Upper Bound

If upper is the upper bound on the cost of a minimum-cost solution, then all live nodes with c(x) > upper may be killed.

- All answer nodes reachable from x have cost c(x) ≥ c(x) > upper.
- The starting value for upper can be obtained by some heuristic or set to 1.
- Each time a new answer node is found, the value of upper can be updated.

Optimization/Minimization Problem

- The maximization is converted to minimization by changing the sign of the objective function.
- Formulate the search for an optimal solution as a search for an LC answer node in a state-space tree.
- Define the cost function c(.) such that c(x) is the minimum for all nodes, representing an optimal solution.

Some examples of branch-and-bound technique are

- the assignment problem,
- the knapsack problem,
- and the traveling salesman problem.

5.7   0/1 KNAPSACK PROBLEM

Given n objects or items with known weight w_i and profit v_i, i = 1, 2, 3, 4, . . . , n and a knapsack or a bag with capacity W, in which subset of items is to be placed. The objective of this problem is to obtain filling the knapsack with the maximum profit earned, but it should not exceed the knapsack capacity W.

> **Solution:** A state-space tree is used to solve the knapsack problem using the branch-and-bound technique.
>
> **Construction of state-space tree:** The state-space tree is constructed as a binary tree in which each node on the ith level $0 \leq i \leq n$ represents all the subsets of n items that include a particular selection made from the first i-ordered items.

This selection is uniquely determined by a path from the root to the node. A branch going to the left indicates that the next item is included. A branch going to the right indicates that the next item is not included. The total weight W and the total value V of this selection in the node are recorded along with some upper bound, Ub, on the value of any subset that can be obtained by adding zero or more items in this selection.

The upper bound Ub is computed as

$$Ub = v + (W - w)(v_{i+1}/w_{i+1}).$$

Here, v = total value of items, W = knapsack capacity, w = the total weight of the items, and v_{i+1}/w_{i+1} = the best payoff per weight unit of the next item.

Example: Consider four items with the following weights and values:

Item	Weight	Profit(v)	v/w
1	7	$49	7
2	3	$12	4
3	6	$18	3
4	5	$30	6

Step 1:
The items are ordered in descending order by their value-to-weight ratio.

Item	Weight	Profit(v)	v/w
1	7	$49	7
4	5	$30	6
2	3	$12	4
3	6	$18	3

Step 2:
The upper bound Ub is computed as

$$Ub = v + (W - w)(v_{i+1}/w_{i+1}).$$

Graph 167

Here, v = total value of items,
W = knapsack capacity,
w = the total weight of the items, and
v_{i+1}/w_{i+1} = the best payoff per weight unit of the next item.

STATE-SPACE TREE OF THE KNAPSACK PROBLEM USING THE BRANCH-AND-BOUND TECHNIQUE

The state-space tree can be constructed as a binary tree like that in Figure 5.27. The root of the tree represents the starting point, with no decisions about the given elements made as yet.

EXPLANATION

1. Construction of node 0 at level 0.
   No items have been selected.

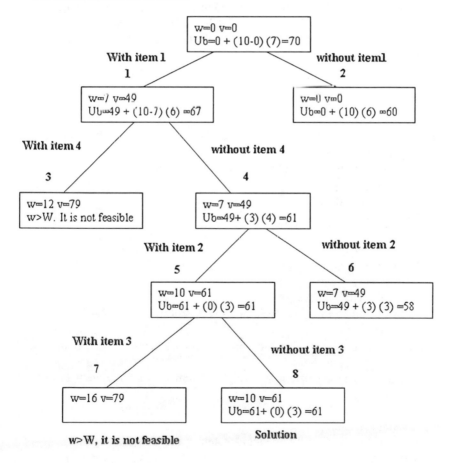

FIGURE 5.27 State-space tree

Consider $w = 0$, $v = 0$

$$v_{i+1}/w_{i+1} = 0$$
$$W = 10$$
$$Ub = v + (W - w)(v_{i+1}/w_{i+1})$$
$$= 0 + (10 - 0)(7) = 70$$

2. Construction of node 1 that includes item 1
   The total weight $w = 7$, $v = 49$, $W = 10$, $v_{i+1}/w_{i+1} = 6$.
   $$Ub = v + (W - w)(v_{i+1}/w_{i+1})$$
   $$= 49 + (10 - 7)(6) = 67$$

3. Construction of node 2 that does not include item 1
   The total weight $w = 0$, $v = 0$, $W = 10$, $v_{i+1}/w_{i+1} = 6$
   $$Ub = v + (W - w)(v_{i+1}/w_{i+1})$$
   $$= 0 + (10 - 0)(6) = 60$$
   Here the upper bound Ub is 60, which is inferior to node 8.
   At this level, node 1 is a promising node, so further calculations are done from that node.

4. Construction of node 3 that includes item 1 and item 4, respectively
   The total weight $w = 7 + 5 = 12$, $v = 79$, $W = 10$, $v_{i+1}/w_{i+1} = 4$
   Since the total weight ($w = 12$) exceeds the knapsack capacity ($W = 10$), it is not feasible.

5. Construction of node 4 with item 1 and without item 4
   The total weight $w = 7$, $v = 49$, $W = 10$, $v_{i+1}/w_{i+1} = 1$
   $$Ub = v + (W - w)(v_{i+1}/w_{i+1})$$
   $$= 49 + (10 - 7)(4) = 61$$

6. Construction of node 5 that includes item 1 and item 2 but not item 4, respectively
   The total weight w = weight of item 1 + weight of item 2
   $$= 7 + 3 = 10$$
   $v = 61$, $W = 10$, $v_{i+1}/w_{i+1} = 3$
   $$Ub = v + (W - w)(v_{i+1}/w_{i+1})$$
   $$= 61 + (10 - 10)(3) = 61$$

7. Construction of node 6 that includes item 1 but not item 4 or item 2
   The total weight $w = 7$, $v = 49$, $W = 10$, $v_{i+1}/w_{i+1} = 3$
   $$Ub = v + (W - w)(v_{i+1}/w_{i+1})$$
   $$= 49 + (10 - 7)(3) = 58$$
   At level 3, the most promising node is 5, so further calculations are done from that node.

8. Construction of node 7, which has the subset items—item1, item 2, and item 3 and without item 4
   Therefore, the total weight w = weight of item 1 + weight of item 2 + weight of item 3
   $$= 7 + 13 + 6 = 16$$, which exceeds the knapsack capacity.
   It is not feasible.

9. Construction of node 8 that includes item 1 and item 2.
   Therefore, the total weight $w = 10$, $v = 61$, $W = 10$, $v_{i+1}/w_{i+1} = 3$

Graph   **169**

$$Ub = v + (W - w)(v_{i+1}/w_{i+1})$$
$$=61 + (10 - 10)(3) = 61$$

There is no next item. So the upper bound is equal to the total value of these items.

FIFO SOLUTION

The FIFO version uses a queue to keep track of live nodes, as these nodes are to be extracted in FIFO order.

Example

Consider the knapsack instance n = 3, w = [20, 15, 15], p = [40, 25, 25], and
C = 30.

The FIFO branch-and-bound search begins with the root A as the E-node. At this time, the live node queue is empty. When node A is expanded, nodes B and C are generated. As both are feasible (w = 20 when item 1 is included, w = 0 when item 1 is not included), they are added to the live-node queue, and node A is discarded.

Live-node queue	B	C	

The next E-node is node B. It is expanded to get nodes D and E. At D, w = 35; at E, w = 20. D is infeasible (w > C, 35 > 30) and discarded, while E is added to the queue.

Queue	B	C	E

Next, C becomes the E-node; when expanded, it leads to nodes F and G. Both are feasible and added to the queue.

Queue	B	C	E	F	G	

The next E-node gets us to J and K. J is infeasible and discarded. K is a feasible leaf and represents a possible solution to the instance. Its profit value is 40.

The next E-node is node F. Its children L and M are generated.

L represents a feasible packing with a profit value of 50, while M represents a feasible packing with a value of 15.

G is the last node to become the E-node. Its children N and O are both feasible. The search now terminates because the live-node queue is empty. The best solution found has a value of 50.

5.8 NP-HARD AND NP-COMPLETE PROBLEMS

BASIC CONCEPTS

- There are algorithms for which there is no known solution. For example, Turing's halting problem.
- The halting problem cannot be solved by any computer no matter how much time is provided.

CLASSIFICATION OF PROBLEMS

The distinction between problems that can be solved by a polynomial time algorithm and problems for which no polynomial time algorithm is known can be classified into one of the two groups:

1. Tractable
2. Intractable

Tractable problem: Problems whose solution times are bounded by a polynomial with a small degree are called tractable algorithms. That is, there exists a polynomial time algorithm that solves the problem. An algorithm is polynomial-bound if its worst-case growth rate can be bound by a polynomial p(n) in the size n of the problem.

$$p(n) = a_n n^k + \ldots\ldots\ldots + a_1 n + a_0, \text{ where } \mathbf{k} \text{ is a constant.}$$

Examples
- Searching **O(logn)**
- Polynomial evaluation **O(n)**
- Sorting **O(nlogn)**
- String editing **O (mn)**

Intractable problems: Problems whose best known algorithms are not bounded by a polynomial, that is, a nonpolynomial, are called hard or intractable problems. All algorithms that solve the problem are not polynomial-bound. It has a worst-case growth rate f (n) that cannot be bound by a polynomial p(n) in the size n of the problem.

For intractable problems, the bounds are

$$f(n) = c^n, \text{ or } n^{\log n}.$$

Examples
- Traveling salesperson problem **O ($n^2 2^n$)**
- Knapsack problems **O ($2^{n/2}$)**

THEORY OF NP-COMPLETENESS

The theory of NP-completeness shows that many of the problems with no polynomial time algorithms are computationally related. The group of problems is further subdivided into two classes:

1. NP-complete
2. NP-hard

P: P is the class of decision problems that can be solved by a deterministic polynomial algorithm.

Graph   171

NP: NP is the class of decision problems that can be solved by a nondeterministic polynomial algorithm.

NP-complete: A problem that is NP-complete has the property that it can be solved in polynomial time if and only if all other NP-complete problems can also be solved in polynomial time.

NP-hard: If an NP-hard problem can be solved in polynomial time, then all NP-complete problems can be solved in polynomial time.

All NP-complete problems are NP-hard, but some NP-hard problems are not known to be NP-complete. The relationship among P, NP, and NPC is shown in Figure 5.28.

NONDETERMINISTIC ALGORITHMS

The algorithm that has the property that the result of every operation is uniquely defined is termed a *deterministic algorithm*. Such algorithms agree with the way programs are executed on a computer.

Nondeterministic algorithms are allowed to contain operations whose outcomes are not uniquely defined but are limited to specified sets of possibilities. The machine executing such operations is allowed to choose any one of these outcomes subject to a termination condition to be defined later. This leads to the concept of a *nondeterministic algorithm*.

This algorithm is specified with the help of three new functions:

1. Choice (S)
   - This function arbitrarily chooses one of the elements of set S.
   - The assignment statement x = Choice (1, n) could result in *x* being assigned any one of the integers in the range **[1, n]**.
   - There is no rule specifying how this choice is to be made.

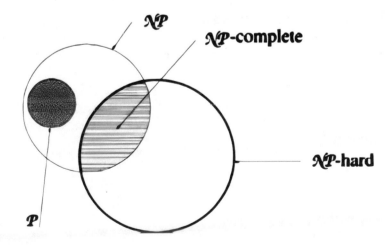

FIGURE 5.28   Relationship among P, NP, and NPC

2. Failure ()
  • This function signals an unsuccessful completion.
  • It cannot be used as a return value.
3. Success ()
  • The function signals a successful completion.
  • It cannot be used as a return value.
  • Whenever there is a set of choices that leads to a successful completion, then one such set of choices is always made and the algorithm terminates successfully.

A nondeterministic algorithm terminates unsuccessfully if and only if there exists no set of choices leading to a success signal. A machine capable of executing a nondeterministic algorithm in this way is called a nondeterministic machine.
A nondeterministic algorithm consists of two phases:

Phase 1: guessing
Phase 2: checking

If the checking stage of a nondeterministic algorithm is of polynomial time complexity, then this algorithm is called an NP (nondeterministic polynomial) algorithm. NP problems must be decision problems.

Example
  • Searching
  • MST
  • Sorting
  • Satisfiability problem
  • Traveling salesperson problem

Nondeterministic Searching Algorithm
Consider the problem of searching for an element x in a given set of elements A[1:n], $n \geq 1$. The index j is such that A[j] = x or j = 0 if x is not in A.

```
Void Nd_Search(A,n,x)
{
        j = Choice (1, n)
        if (A[j] = = x)
        {
        cout <<j;
        success();
        }
        Cout<<0;
        Failure();
}
```

Graph   **173**

By the definition of nondeterministic algorithm, the output is 0 iff there is no j such that A[j] = x. Since A is not ordered, every nondeterministic algorithm has a complexity of O(1).

Note that since *A* is not ordered, every deterministic search algorithm is of complexity O(n).

Nondeterministic Sorting Algorithm

Consider the problem of sorting an array of positive integers, A[i], $1 \leq i \leq n$ in non-decreasing order.

```
void NSort(int A[], int n)
// Sort n positive integers.
{
        int B[SIZE], i, j;
        for (i=1; i<=n; i++) B[i]=0; // Initialize B[].
        for (i=1; i<=n; i++)
        {
                j = Choice(1, n);
                if (B[j]) Failure();
                B [j] = A [i] ;
        }
        for (i=1; i<=n-1; i++) // Verify order.
        if (B[i] > B[i+1]) Failure();
        for (i=1; i<=n; i++) cout « B [i] « J ';
        cout « endl; Success();
}
```

- A deterministic interpretation of a nondeterministic algorithm can be made by allowing unbounded parallelism in computation.
- Each time a choice is to be made, the algorithm makes several copies of itself.
- One copy is made for each of the possible choices. Thus, many copies are executed at the same time.
- The first copy to reach a successful completion terminates all other computations. If a copy reaches a failure completion, then only that copy of the algorithm terminates.
- In case there is no sequence of choices leading to a successful termination, we assume that the algorithm terminates in one unit of time with output 'unsuccessful computation'.
- Whenever a successful termination is possible, a nondeterministic machine makes a sequence of choices that is the shortest sequence, leading to a successful termination.

Definition: Any problem for which the answer is either zero or one is called a ***decision problem***. An algorithm for a decision problem is termed a ***decision algorithm***.

Any problem that involves the identification of an optimal (either minimum or maximum) value of a given cost function is known as an *optimization problem*. An *optimization algorithm* is used to solve an optimization problem.

Examples

Maximum Clique

A maximal complete subgraph of a graph $G = (V, E)$ is a *clique*. The size of the clique is the number of vertices in it. The *max clique problem* is an optimization problem that has to determine the size of a largest clique in G. The corresponding decision problem is to determine whether G has a clique of size at least k for some given k.

Let DClique(G, k) be a deterministic decision algorithm for the clique decision problem. If the number of vertices in G is n, the size of a max clique in G can be found by making several applications of DClique. DClique is used once for each k, $k = n, n - 1, n - 2 \ldots$ until the output from DClique is 1.

If the time complexity of DClique is f(n), then the size of a max clique can be found in time $g(n) \geq nf(n)$. Also, if the size of a max clique can be determined in time g(n), then the decision problem can be solved in time g(n). Hence, the max clique problem can be solved in polynomial time if and only if the clique decision problem can be solved in polynomial time.

0/1 Knapsack

The knapsack decision problem is to determine whether there is a 0/1 assignment of values to X_i, $1 \leq i \leq n$ such that $\sum P_i X_i \geq r$ and $\sum W_i X_i \geq m$ for given **m** and **r**. The r is a given number. The **Pi's** and **Wi's** are nonnegative numbers. If the knapsack decision problem cannot be solved in deterministic polynomial time, then the optimization problem cannot either.

Uniform Parameter n to Measure Complexity

- Assume that n is the length of the input to the algorithm (that is, n is the input size) and assume that all inputs are integers.
- Rational inputs can be provided by specifying pairs of integers.
- Generally, the length of an input is measured assuming a binary representation; that is, if the number 10 is to be input, then in binary, it is represented as 1010. Its length is 4.
- A positive integer **k** has a length of $\log_2 k + 1$ bits represented in binary. The length of the binary representation of 0_2 is 1.
- The size, or length, n of the input to an algorithm is the sum of the lengths of the individual numbers being input.
- When inputs are given using the radix **r = 1**, we say the input is in *unary form*. In unary form, the number 5 is input as 11111. Thus, the length of a positive integer **k** is **k**.
- It is important to observe that the length of a unary input is exponentially related to the length of the corresponding r-ary input for radix **r, r > 1**.

Graph 175

Max Clique

The input to the max clique decision problem can be provided as a sequence of edges and an integer **k**. Each edge in **E(G)** is a pair of numbers **(i, j)**. The size of the input for each edge **(i, j)** in binary representation is $log_2 i + log_2 j + 2$.

The input size of any instance is

$$n = \sum_{i,j \in E(G)} \left(log_2 i + log_2 j + 2 \right) + log_2 k + 1.$$

Note that if G has only one connected component, then $n \geq |V|$. Thus, if this decision problem cannot be solved by an algorithm of complexity **p(n)** for some polynomial **p()**, then it cannot be solved by an algorithm of complexity $p(|V|)$.

0/1 Knapsack

The input size q for the knapsack decision problem is

$$q = \sum_{1 \leq i \leq n} \left(\lfloor log_2 P_i \rfloor + \lfloor log_2 W_i \rfloor \right) + 2n + \lfloor log_2 m \rfloor + \lfloor log_2 r \rfloor + 2.$$

If the input is given in unary notation, then input size $s = \sum P_i + \sum W_i + m + r$.

Knapsack decision and optimization problems can be solved by an algorithm of complexity **p(s)** for some polynomial **p()**.

Definition: Complexity of Nondeterministic Algorithm

The *time required by a nondeterministic algorithm* performing on any given input is the minimum number of steps needed to reach a successful completion if a sequence of choices leading to such a completion exists.

In the case that successful completion is not possible, then the time required is **O(1)**. A nondeterministic algorithm is of complexity **O(f(n))** if for all inputs of size $n, n \geq n_0$, that result in a successful completion, the time required is at most $cf(n)$ for some constants c and n_0.

In the earlier definition, we assume that each computation step is of a fixed cost. In word-oriented computers, this is guaranteed by the finiteness of each word. When each step is not of a fixed cost, it is necessary to consider the cost of individual instructions. Thus, the addition of two m-bit numbers takes **O(m)** time, their multiplication takes **O(m²)** time (using classical multiplication), and so on.

Deterministic Decision Algorithm to Get a Sum of Subsets

Consider the deterministic decision algorithm to get a sum of subsets.

```
void Sum(int A[], int n, int m)
{
long unsigned int s = 1;
//s is an (m + l)-bit word. Bit zero is 1.
for (int i=l; i<=n; i++)
```

```
s = s I (s << A[i]); //<< is the left shift operator.
int t = (s>> m);
if ((1 I t) == t) // If the mth bit of s is 1
cout « "A subset sums to m." « endl;
else cout « "No subset sums to m." « endl;
}
```

- Consider the deterministic decision algorithm from 0 to m from right to left
- Bit **i** will be **0** if and only if no subsets of **A[j]**, $1 \leq j \leq$ sums to **i**.
- Bit **0** is always 1, and bits are numbered 0, 1, 2, . . . , m right to left.
- The number of steps for this algorithm is **O(n)**.
- Each step moves m +1 bits of data and would take **O(m)** time on a conventional computer.
- Assuming one unit of time for each basic operation for a fixed word size, the complexity of deterministic algorithm is **O(nm)**.

Knapsack Decision Problem

The nondeterministic polynomial time algorithm for the knapsack problem follows:

```
void DKP(int p[], int w[], int n, int m, int r, int x[])
{       int W=0, P=0;
        for (int i=1; i<=n; i++)
          {
          X[i] = Choice(0,l);
          W+= x[i]*w[i]; P += x[i]*p[i];
          }
        if ((W > m) ||(P < r)) Failure();
        else Success();
}
```

- The *for* loop selects or discards each of the **n** items.
- It also recomputes the total weight and profit corresponding to the selection.
- The *if* statement checks to see the feasibility of assignment and whether the profit is above a lower bound **r**.
- The time complexity of the algorithm is **O(n)**.
- If the input length is **q** in binary, the time complexity is **O(q)**.

Nondeterministic Decision Algorithm for a Clique Decision Problem

The nondeterministic polynomial time algorithm for the clique decision problem follows:

```
void DCK(int G[] [SIZE], int n, int k)
{
        S = 0; // S is an initially empty set.
        for (int i=1; i<=k; i++)
```

Graph   177

```
        {
        int t = Choice(1, n);
        if (t is in S) Failure();
        S = S U {t}; // Add t to set S.
        }
        //At this point S contains k distinct vertex indices.
        for (all pairs (i,j) such that i is in S,
        j is in S and i!=j)
        if ((i,j) is not an edge of G) Failure();
        Success();
}
```

The time complexity of the algorithm is $O(n + k^2) = O(n^2) = O(m)$.

SATISFIABILITY

Let $x_1, x_2, \ldots$ denote a set of Boolean variables and denote the complement of $\mathbf{X_i}$.

Literals: A variable or its complement is called a literal.
Formula: A formula in propositional calculus is an expression that is constructed by connecting literals using the operations **and ( )** and **or( )**.

Examples

$$\left(x_1 \wedge \overline{x_2}\right) \vee \left(x_3 \wedge \overline{x_4}\right)$$

$$\left(x_3 \vee \overline{x_4}\right) \wedge \left(x_1 \wedge \overline{x_2}\right)$$

Conjunctive Normal Form

A formula is in conjunctive normal form (CNF) if it is represented as $\wedge^k_{i=1} c_i$ where c_i are clauses represented as $\vee l_{ij}$ where l_{ij} *are* literals.

Examples: $\left(x_3 \vee \overline{x_4}\right) \wedge \left(x_1 \wedge \overline{x_2}\right)$

Disjunctive Normal Form

A formula is in disjunctive normal form (DNF) iff it is represented as $\vee^k_{i=1} c_i$, where c_i are clauses represented as $\wedge l_{ij}$, where l_{ij} *are* literals.

Example: $\left(x_1 \wedge \overline{x_2}\right) \vee \left(x_3 \wedge \overline{x_4}\right)$

Satisfiability problem: A satisfiability problem is to determine whether a formula is true for some assignment of truth values to the variables. CNF-satisfiability is the satisfiability problem for CNF formulas.

A polynomial time nondeterministic algorithm that terminates successfully iff a given propositional formula $E(x_1, \ldots, x_n)$ is satisfiable that is chosen nondeterministically one of the 2^n possible assignments of truth values to $(x_1, \ldots, x_n)$ and verify that $E(x_1, \ldots, xn)$ is true for that assignment.

```
void Eval(cnf E, int n)
//I Determine whether the prop. formula E is satisfiable.
//The variables are x[i], x[2], . . . , x[n].
{
    int x[SIZE];
    //Choose a truth value assignment.
    for (int i=i; i<=n; i++)
    x[i] = Choice(0,l);
    if (E(x,n)) Success();
    else Failure 0;
}
```

- The nondeterministic time to choose the truth value is $O(n)$.
- The deterministic evaluation of the assignment is also done in $O(n)$ time.

THE CLASSES NP-HARD AND NP-COMPLETE

Polynomial complexity: An algorithm **A** is of polynomial complexity if there exists a polynomial p() such that the computation time of A is $\mathbf{O(p(n))}$ for every input of size **n**.

P: P is the set of all decision problems solvable by deterministic algorithms in
a polynomial.
NP: NP is the set of all decision problems solvable by nondeterministic algorithms in polynomial time.

Since deterministic algorithms are a special case of nondeterministic algorithms, $P \subseteq NP$. An unsolved problem in computer science is 'Is P = NP or is $P \neq Np$?'

Cook formulated the following question: Is there any single problem in NP such that if we showed it to be in P, then that would imply that P = NP? This led to Cook's theorem as follows:

Cook's Theorem: Satisfiability is in P if and only if P = NP.
Reducibility:

Let L_1 and L_2 be problems. Problem L_1 reduces to L_2 (written as $L_1 \infty L_2$) if and only if there is a way to solve L_1 by a deterministic polynomial time algorithm using a deterministic algorithm that solves L_2 in polynomial time.

If we have a polynomial time algorithm for L_2, then we can solve L_1 in polynomial time. Reducibility is transitive if

Graph   179

$$L_1 \infty L_2 \quad and \quad L_2 \infty L_3 \ then \ L_1 \infty L_3.$$

- Show that one problem is no harder or no easier than another, even when both problems are decision problems.
- Consider a decision problem **A** which we would like to solve in polynomial time.
- Instance of the problem:
  - Input to the particular problem at hand
  - Given a graph G, vertices **u** and **v**, and an integer **k**, determine if a path exists from **u** to **v** consisting of at most **k** edges.
- Consider a different decision problem B that we already know how to solve in polynomial time.
- Suppose that we have a deterministic procedure that transforms an instance α of **A** to an instance β of B with the following characteristics.
  - Transformation takes polynomial time.
  - The answers are the same; answer for α is 'yes' if and only if the answer to β is also 'yes'.
- It is written as A ≤ B.

NP-Hard

A problem L is NP-hard if and only if satisfiability reduces to L (satisfiability ∞ L). A problem L is NP-complete if and only if L is NP-hard and L ∈ NP is shown in Figure 5.29.

- There are NP-hard problems that are not NP-complete.
- Only a decision problem can be NP-complete.
- An optimization problem may be NP-hard; it cannot be NP-complete.
- If L_1 is a decision problem and L_2 is an optimization problem, it is quite possible that $L_1 \infty L_2$.
  - A knapsack decision problem can be reduced to the knapsack optimization problem.
  - A clique decision problem reduces to the clique optimization problem.

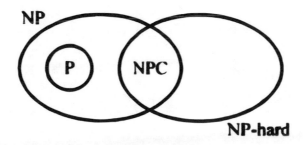

FIGURE 5.29 Relationship among P, NP, NP-complete and NP-hard problems

Halting Problem for Deterministic Algorithms

The halting problem for deterministic algorithms is an example of NP-hard decision problems that are not NP-complete. The halting problem is used to determine for an arbitrary deterministic algorithm A and input I whether A with input I ever terminates, which is undecidable. Hence, there exists no algorithm of any complexity to solve the halting problem. It clearly cannot be NP. To show the satisfiability of the halting problem, construct an algorithm **A** whose input is a propositional formula **X**.

- If X has n variables, A tries out all the 2n possible truth assignments and verifies whether X is satisfiable.
- If X is satisfiable, it stops; otherwise, A enters an infinite loop. Hence, A halts on input X iff X is satisfiable.

If we had a polynomial time algorithm for the halting problem, then we could solve the satisfiability problem in polynomial time using A and X as input to the algorithm for the halting problem. Hence, the halting problem is an NP-hard problem that is not NP.

Polynomially Equivalent

Two problems L_1 and L_2 are said to be polynomially equivalent if and only if $L_1 \propto L_2$ and $L_2 \propto L_1$

- To show that a problem L_2 is NP-hard, it is adequate to show that $L_1 \propto L_2$, where L_1 is some problem already known to be NP-hard.
- Since $\propto$ is a transitive relation, it follows that if satisfiability $\propto L_1$ and $L_1 \propto L_2$, then satisfiability $\propto L_2$.
- To show that an NP-hard decision problem is NP-complete, we have just to exhibit a polynomial time nondeterministic algorithm for it.

Index

adjacency 82, 84, 124, 127, 137–141
algorithmics 1
articulation point 156–161

biconnected graph 157
big-oh notation 20, 21, 24, 26
big-omega notation 20, 22, 26
big-theta notation 20, 22, 23, 26
breadth-first search 112, 119, 139, 140, 162

chromatic number 123
combinatorial problems 10, 11
computational procedures 2
consecutive integer checking algorithm 2, 3

data space 14
debugging 10
definiteness 2
degree 29, 114, 123, 137, 170
depth-first search 109, 111, 116, 139
directed graph 76, 77, 82, 83, 103, 135–137, 140, 143, 146, 148, 159
divide-and-conquer 43–45, 47–49, 57, 61
dynamic programming 7, 75, 76, 83, 89, 104

effectiveness 2
Euclid's algorithm 2, 3
explicit constraints 110, 111

finiteness 2, 175
flowchart 7, 8

generality 9
geometric problems 10, 12
graph problems 10, 11

implicit constraints 110–112
input 1–3, 6–10, 14, 24, 27, 28, 34, 35, 52, 64, 174–179, 180
instruction space 14
intractable 170
iteration method 32

Kruskal's algorithm 148, 153–155

little-oh notation 20, 23
little omega notation 20, 23

master method 32, 33
merge sort 30, 48, 60–64
middle school procedure 2, 4

output 1–3, 7, 10, 14, 18, 35, 110, 128, 173

path 7, 55, 66, 75–89, 103, 104, 107, 111, 118, 133, 143, 149, 160, 164, 166, 179
planar graph 124
polynomial complexity 77, 178
Prim's algorithm 148, 151–153
profiling 10
program proving 10
program step 16, 17
program verification 10
pseudo code 3, 7, 37, 78, 80, 161

searching 10, 11, 36, 87, 109, 170, 172
simplicity 9, 28, 30
sorting 10, 11, 24, 48, 60, 61, 70, 153, 170, 172, 173
space efficiency 9
spanning tree 66, 137, 143, 146, 148, 151, 154–156, 159, 160
stack 13, 14, 112, 162
state-space tree 109, 111, 114, 116, 129, 131, 132, 164–167
string processing 10, 11
substitution method 31, 40, 45

time efficiency 9
tractable 170

undirected graph 135–137, 140, 146, 148, 158, 159
using natural language 7